Defence Industrial Base-2025

Defence Industrial Base-2025

Based on Study by CENJOWS

Contributors

Prof. S.K.Palhan

Mr. H.C.Gandhi & Brig (Ret d) S.Bhalla

Published in association with

Centre for Joint Warfare Studies

New Delhi

Vij Books India Pvt Ltd

21 Ansari Road, Daryaganj, New Delhi

Published by

Vij Books India Pvt Ltd
(Publishers, Distributors & Importers)
4675-A, 21 Ansari Road
New Delhi - 110002
Phones: 91-11-43596460, 91-11- 65449971
Fax: 91-11-30126465
e-mail : vijbooks@rediffmail.com

ISBN: 978-93-80177-34-2

CONTENTS

Foreword

India is aspiring to occupy a place on the high table of nations, it is being hailed as a Regional power already The leadership of the country is aiming to usher in prosperity and security to its citizens. It is axiomatic that strong economic growth requires sustained peace which calls for strong defence capability to meet the external threats and act as deterrence.

Revolution in Military affairs has made armaments/weapons a major component of defence capability. The expotential growth of technology has resulted in the cutting edge being eroded very rapidly necessitating constant up gradation/replacement of defence hardware.

As a regional power, it thus becomes mandatory for India to have it s defence forces equipped with the latest technology so as to have an edge over the adversaries. This may not be possible if we continue to import 70% of our defence requirement. The imports only allow obsolete or near obsolete technologies and are never the state of the art. For balance 30% of indigenous capability heavy reliance, as of now has been placed on Defence PSUs with little in house R&D capability. It therefore becomes paramount that we be self reliant by creating a sound Defence Industrial Base (DIB) - with a wider base to utilise our nation's industrial potential.

The framework of options to strategically and realistically absorb the offsets to strengthen our Defence Industrial Base requires a comprehensive study to aim at the long term objective of a sound DIB by 2025.

Apart from Government policies and capital intensiveness the DIB also has to view production and technology management procedures, research and development base, and human resource development for its sustenance.

Considering these imperatives, Centre for Joint Warfare Studies undertook to carryout a study of Indian Defence Industrial Base 2025. This is a pioneering effort in this field which has been done by the study group headed by Prof SK Palhan - a seminal work on the subject.

The book examines the technological life cycle of defence equipment, and enumerates the experiences and practices followed by some developed countries towards creating a vibrant defence industrial base. This is followed by an analysis of the policies with regard to Defence Expenditure, Defence Acquisition and Offsets clauses for import.

The present status of Indian DIB, Research and Development Infrastructure, Ordnance factories and other Science and Technology agencies too has been elucidated to arrive at what is presently available.

The sectoral requirements for future in terms of capability required by our land, sea and air forces have also been identified to include the Electronic and Cyber Warfare domain.

With the collation of all the relevant facts, recommendations to achieve the required DIB have been listed to cover Policy frame work, Technology Management, Production, Human Resources Development and Research and Development. These include both short term and long term measures with distinct time lines. The role of private industry has also been enunciated in the book as with coming of age, the industry must find expression in defence sector also.

The book, it is hoped, would help the Policy makers, Research and Development agencies and industry to have an insight into the evolution of Defence Industrial Base. It may be a step towards achieving the requisite self reliance for India and self sustenance of the armed forces.

The team has made a sincere effort to cover all relevant aspects on the subject yet, considering the complexity of the subject and to maintain a continuous flow of information, suggestions are welcome.

25 Jun 2010
New Delhi

Maj Gen K B Kapoor
Director, CENJOWS

Acknowledgement

We are grateful to Air Marshal S.C.Mukul, AVSM, VM, VSM, CISC & Chairman CENJOWS for supporting this study. We are grateful to Lt. Gen. (Retd) H.S Lidder, PVSM, UYSM, YSM, VSM, ADC, Former CISC for providing the broad vision for developing the Defence Industrial Base of the country by 2025. The guidance give in the two review meetings held under his chairmanship were very valuable. Maj. Gen. (Retd) K.B. Kapoor, VSM, Executive Director CENJOWS helped us in understanding of the Defence requirements through his vast experience. We would like to acknowledge the experts from HQ IDS for elaborating the futuristic need of the services. Rear Adm. Chakravarty gave critical inputs for direction of the study

The study could not have been completed without the valuable inputs from Mr.S.Radhakrishnan from DRDO, DGQA-special, DGOF, and DG EME. For the data on R&D the inputs from Dr. Laxman Prasad Advisor DST were very useful. Dr. Bhojwani Advisor to the Minister of Science & Technology & Dr. Naresh Kumar Head, R&D Planning Division CSIR gave the background data on the CSIR resources. For the data in the private sector we are grateful to CII and FICCI for the valuable contribution. Prof. S.C Jain Dean R & D, IIT Delhi gave very useful suggestions for enlarging the role of academic institutions in building the basic Research and Manpower component of the DIB.

Col. A.K. Singh, Col. Mann & Col. Tejbir Singh from CENJOWS / IDS provided all the support for carrying out the study. We would like to acknowledge the valuable contribution made by the experts in the brainstorming sessions on the study

Prof. S.K.Palhan
Advisor CENJOWS
(Chairman study Team)
Member- Mr. H.C Gandhi
Member- Brig S. Bhalla ((Retd)

ABBREVIATIONS

ABM	Anti Ballistic Missile
AOC	Army Ordnance Corps
ASAT	Anti Satellite
BEL	Bharat Electronics Limited
BIS	Bureau of Industry and Security
BMDS	Ballistic Missile Defence System
BMEW	Ballistic Missile Early Warning
BPR	Business Process Review
CAG	Comptroller and Auditor General
CENJOWS	Centre for Joint Warfare Studies
CIDS	Chief of Integrated Defence Staff (India)
CLAWS	Centre for Land Warfare Studies
COCOM	Coordinating Committee for Mutual Export Control
CSIR	Council of Scientific and Industrial Research
DAC	Defence Acquisition Council
DAE	Department of Atomic Energy
DARPA	Defence Advanced Research Project Agency
DGOF	Director General Ordnance Factories
DIB	Defence Industrial Base
DOFA	Defence Offset Facilitating Agency
DPB	Defence Procurement Board
DPM	Defence Procurement Manual (for revenue expenditure)
DPASS	Defence Priority and Allocation System Program
DPSU	Defence Public Sector Undertaking
DRDO	Defence Research and Development Organisation
DST	Department of Science and Technology
EAR	Export Armament Regulations
EEZ	Exclusive Economic Zone
EME	Electrical and Mechanical Engineers
FDI	Foreign Direct Investment
FELIN	Fentassin Equipments et Liaison Integers
FINSAS	Future Infantry Soldier System
FIST	Future Integrated Soldier System
FOL	Fuel Oil and Lubricants

GBI	Ground Based Interceptor
GDP	Gross Domestic Product
GPS	Global Positioning System
GSLV	Geosynchronous Satellite Launch Vehicle
HAL	Hindustan Aeronautics Limited
ICAR	Indian Council of Agricultural Research
ICBM	Intercontinental Ballistic Missile
IDS	Integrated Defence Staff
IGMDP	Integrated Guided Missile Development Program
INSAT	Indian Satellite
IPR	Intellectual Property Rights
IRBM	Intermediate Ballistic Missile
IRS	Indian Remote Sensing Satellite
ISRO	Indian Space Research Organisation
ITAR	International Traffic in Arms Regulation
JMSDF	Japan Maritime Self Defence Forces
LRTR	Long Range Tracking Radar
LTIPP	Long Term Integrated Perspective Plan
MD	Missile Defence
MEADS	Medium Extended Air Defence System
MRCA	Multi Role Combat Aircraft
MTCR	Missile Control Technology Regime
NASA	National Aeronautic and Space Administration
NATO	North Atlantic Treaty Organization
NBC	Nuclear Biological and Chemical
NCNC	No Cost No Commitment
NCW	Network Centric Warfare
NDU	National Defence University
NKC	National Knowledge Commission
OEM	Original Equipment Manufacturer
OFB	Ordnance Factory Board
PLA	People Liberation Army
PSU	Public Sector Undertaking
QR	Qualitative Requirement
R&T	Research & Technology (Military Technology)
RDT&E	Research Development Testing & Evaluation
RFP	Request for Proposal
RMA	Revolution in Military Affairs

RUR	Rashtriya Udyog Rattan
SALT	Strategic Arms Limitation Talks
SCM	Supply Chain Management
SDI	Strategic Defence Initiative
SME	Small & Medium Enterprise
TOT	Transfer of Technology
TQM	Total Quality Management
UAV	Unmanned Aerial Vehicle
WMD	Weapon of Mass Destruction
TMC	Tactical Missile Corporation of Russia

Chapter 1
Defence Industry and Trade

Introduction

1. Self reliance in defence requirement s is one of the aims of the Ministry of Defence. India is an emerging power in SouthAsia and needs to protect its assets in the country as well as offshore. It has to protect the lanes of shipping (SLOC) as well as routes of energy Post independence India' s Industrial policy of 1956 dict ated the industrial production, wherein defense, Atomic energy and Railways were reserved for the Public sector . The industrial policy was deeply influenced by the soviet pattern. Prof Mahanoblis the architect of the policy was assisted by three experts deputed from Moscow The defense production is being carried out in the public sector in 39 DGOF factories and 8 DPSUs'. Defense R&D is the exclusive domain of the DRDO. India's defense requirements are imported 70% and the indigenous content is 30 % only of which about 40% (12% of the total) is contributed by the SME of the private sector.

2. The heavy preponderance on import poses tremendous uncert ainty in the defense capability as well as high cost of maintenance and equipment exploitation. This dependence on imports affects the foreign policy and therefore the trade and commerce of India. As a growing economic power India therefore needs to exercise an independent policy to ensure the growth and development of it s economy and enhance the welfare measures for it s people. It is therefore logical that the current proportion of import s be reversed with the indigenous content and steadily reduced further.

3. Dependence on imports directly effects national security capability as demonstrated in the Aircraft carrier Gorshkov case, the supply of nuclear fuel to India and Russian backtracking on the rocket engine technology transfer to India by Russia under pressure from USA. The imperative for economic development largely hinges on developing the Indian DIB. Purely in the milit ary context DIB af fects the cap ability to fight and cont ain terrorism internally and force projection externally not only to counter the growing presence of foreign powers in the Indian Ocean and areas contingent to India but also safeguard offshore natural assets.

4. The next question is what should be the quantum of the DIB and the quantum required by the Indian defense forces, since any industry grows or stagnates depending upon the market size. The market size for defense products is limited to the size of the defense forces and hence is limited, unless products are exported. Therefore it is necessary that as much as possible technology and products available in the civil sector be utilized as also dual use technology developed. This would mitigate the problem of economy of scale in the DIB which can then be enhanced by the export market. Export of defense product are normally carried out of older models for obvious reasons. This highlights the need for constant up gradation of products and technology and importance of R&D.

5. In any country the private sector industry enjoys distinct advantages over the public sector by way of motivation, decision making, organizational control, investment and marketing. These need to be exploited. In the Indian context the private sector lost a lead-time of almost 50 years post independence which now has to be covered. Fortunately the current industrial capability, technical know how and the world wide recession presents a unique opportunity to be seized and exploited.

6. The benefits of having an indigenous DIB are obvious but to summarise the same the following are the benefits:-

- Independence of defense preparedness and reduced dependence on foreign countries.
- Development of indigenous technology. This technology may have dual use and provides an edge over other countries.
- Better bargaining power in international markets to obtain technology available with them in exchange of our technology.
- Military edge over adversaries is ensured by deploying better technology weapons.
- Saving in foreign exchange by reducing imports of military hardware.
- Development of talent within the country and reversing 'Brain-drain' to 'Brain-gain'.
- Generation of employment by establishing production lines in the country of new products and services in the defence sector.
- Development of Industry and trade and general economic conditions.
- Export earnings by way of sale of military hardware to other countries.
- National Pride developed by developing and deploying frontier technology.
- Development of science & technology in the country resulting in multiple advantages.

7. India has developed a base for indigenous manufacturer of defence products through Ordnance factories and the Defence Public sector undertakings. The requirements of the defence forces are dynamic in nature depending on the threat perceptions. Internationally the technology for the defence armaments is progressing at a very fast pace. Since 2001 the resolve has been to reverse the import : indigenous ratio from 70:30 to 30:70 unsuccessfully, since the policy and infrastructure were not in place.

8. The Defence production system has so far prevailed in an environment of monopoly buyer and seller situation. This is not conducive for competitive production. There has been no competition among the Ordnance factories or the Defence PSUs and hardly any competition from the private sector. Thus the products manufactured are not competitive in terms of quality , price and delivery . For developing st ate of the art defence industry which can withst and international competition to meet the domestic demand as well as exports, major changes have to be implemented. The Government is aware of this and a Special Board has been set up with represent atives from Ministry of Defence, Army, and DRDO for providing inputs for planning and upgrading technology of products and efficient functioning of Ordnance Factory Board.Therefore one of the major issues is the up gradation of system for indigenous production of armaments.

9. Considering the international scenario, the procurement of technology for the strategic equipment is not easy due to the various restrictions India has faced during the last three decades. India was denied the import of dual use technologies by COCOM (Coordinating Committee for Multilateral Export Control) up to 1994. Subsequently the denial continues under the Wassenaar Arrangement and MCTR (Missile Control Technology Regime).Even when the technology is available, it is invariably outdated and its supply may be restricted or stopped due to political considerations. It is observed that imported technology invariably involves the import of critical component s and sp ares which imply perpetual external dependence .Technology is not absorbed fully in most of the cases and over a period of time we need further import of technology.

10. Science and technology have been the prime powers behind the human progress. Even a cursory glance at the international scene clearly brings about the crucial role of technology in the growth of different economies. In fact, the critical importance of technology in accelerating the pace of economic development and modernization of traditional societies is now well established. In addition to economic development, technology is also essential for the security and self reliance of a country.

11. Technology has been defined in different ways by economists and social scientists. In the defence parlance it is referred to as Revolution in Milit ary Affairs (RMA).In concrete

terms industrial technology is the knowledge of details of process, design engineering, selection of inputs, optimization of operational and manufacturing techniques, application procedures and quality control as a holistic package for manufacture of an item or delivery of services. The technology package consists of various components. Technology has a cost, a life cycle, a long gestation period and it is transferable and marketable. In general the sale and transfer of defence technology takes place in an imperfect market and price is negotiable

12. State of the art technology is rarely transferred internationally. Bulk of technology transfer takes place when new / superior technology has been developed by the supplier to keep his supremacy over the recipient of technology. The present system of technology transfer may facilitate the expansion of output in the developing countries but this has persistently reinforced technological dependence of countries (UNIDO-1981).The route to technological self reliance is through Indigenous research and development. Leadership in technology is not possible without basic scientific Research and development.

13. A developing country that wishes to accelerate its Defence Industrial base, broadly has two options as under:-

(a) Develop the technology indigenously to meet its goals.

(b) Acquire the technology from abroad.

14. The first option of indigenous development of technology requires scientific / technical manpower, R&D infrastructure and finance. It definitely calls for patience to develop technology which may take its own time. Most of the developing countries including India did not have adequate infrastructure to develop and modernize their technologies and hence depended on other countries for import of technology. In the international scene there have been exceptions where countries opted for technology development within their country for some period. Technology cutoffs can have an exceptional stimulating effect on indigenous capability. As the Russians reduced technical aid and assistance to China in fifties, China concentrated more on indigenous development of technology and virtually had no formal import of technologies from other countries from 1950 to 1979. Incidentally India also developed technologies in the area of Nuclear, space and computer technologies in spite of technology denial atmosphere especially after the Pokhran nuclear test in 1974. The time frame for technology innovation to its use in the field is very long and in many cases it may be superseded by new developments abroad.

15. The second option for import of technology in the defence sector also has many limitations. Firstly the technology which the donors agree to transfer is invariably outdated,

secondly the price is high due to low bargaining power of the recipient, thirdly the technology is not absorbed due to various reasons. Import of technology also has the potential to promote further imports (dealers & vested interests) and kill the indigenous initiative.

16. Technology has a life cycle. New technologies are developed, and after sometime they become less relevant when new/ competitive technologies are developed and in some cases the older technologies are phased out. The different phases in the life cycle of technology may be identified as follows:-

- Innovation stage when the new technology has been developed by individuals / R&D organizations (this may be patented or kept as secret).
- Early adoption of technology by Visionaries who see potential in the technology which is yet to be proven in the field.
- Early maturity stage where the technology is well established and used by Pragmatists.
- Late maturity is the stage in which the conservatives also adopt the technology. The technology at this stage becomes similar to a commodity with number of suppliers.
- Laggards stage where technology has been phased out due to further innovations.

17. It is therefore logical that investment in developing technology must coincide with the technology development life cycle where focus is in the first quadrant breakthrough in which will lead to leadership and early advantage to the user, besides it has a spin off effect of imparting bargaining power.

18. Technology adoption life cycle has been explained in the diagram given below :-

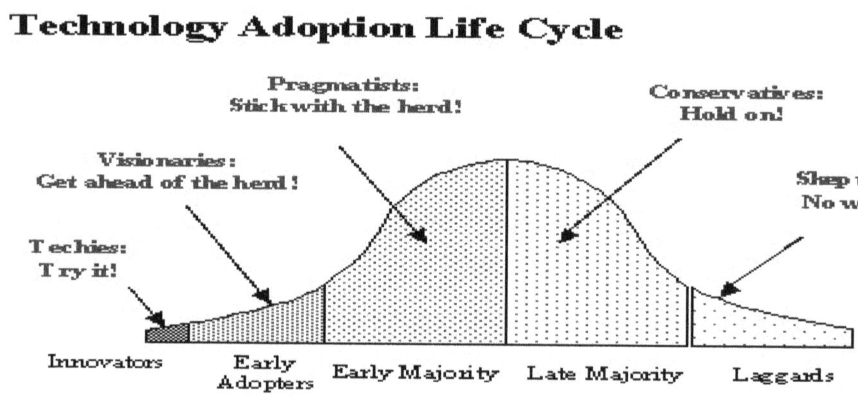

Technology –A when it reaches the decline st age, may be replaced by innovative Technology-B. Technology A is given for transfer at a time when replacement technology B has been developed as shown below.

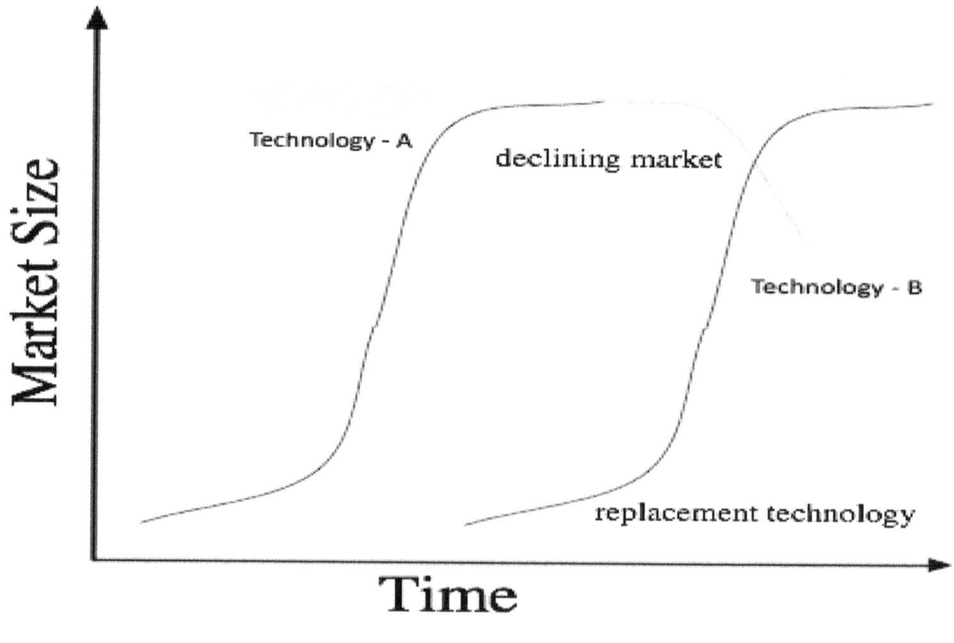

Acquisition of Technology

19. Acquisition of technology from abroad may be done through different channels. Different countries have used different channels at different times depending upon their st age of stage of development, financial strength and policies. Broadly speaking the technologies may be acquired through two channels.

20. Formal channels - These are invariably commercial transactions characterized by signing of agreements. Major formal channels are foreign aid, foreign loans, turnkey project, foreign investment, technical and /or financial collaboration, foreign technicians, consultants, training, engineering consultancy, import of drawing and designs, patents licensing, .Co-production and joint R&D are also formal channels . Import of capit al goods is also an important channel for technology transfer. It may not have a formal agreement on technology but invariably, process or production technology is often built into plant and equipments. Informal channels for technology transfer are mobility of manpower , technical literature, journals, technical seminars, raw material suppliers, and reverse engineering. Most of the countries use mix of both the channels.

Denial of Dual use Technology to India

21. India was denied the import of dual use technologies by COCOM (Coordinating Committee for Multilateral Export Control) up to 1994. Later on the denial is continued under the Wassenaar Arrangement and MTCR (Missile Technology Control Regime).This has been one of the reasons for delays in many R&D projects.

22. Post Cold War, members of the former COCOM export control regime recognized that COCOM's East-West focus was no longer the appropriate basis for export controls.There was a need to establish a new arrangement to deal with risks to regional and international security and stability related to the spread of conventional weapons and dual-use goods and technologies. Accordingly, on the 16th of November 1993, in The Hague, at a High Level Meeting (HLM), representatives of the 17 COCOM member states agreed to terminate COCOM, and establish a new multilateral arrangement, temporarily known as the "New Forum".

23. This decision was confirmed at a further high level meeting in Wassenaar, Netherlands on 29-30 March 1994. COCOM ceased to exist after March 31, 1994. Participating States also agreed to continue the use of the COCOM control lists as a basis for global export controls on a national level until the new arrangement could be est ablished. At this time the former COCOM cooperating countries, namely Austria, Finland, Ireland, New Zealand, Sweden and Switzerland, were included as participating states in the "New Forum". The Russian Federation, Czech Republic, Hungary , Poland, and the Slovak Republic were

welcomed as participating states at the High Level Meeting held on 11-12 September 1995 in Wassenaar. With this major milestone accomplished, the Working Groups were urged to expedite their work. The Participating States of the Wassenaar Arrangement are: Argentina, Australia, Austria, Belgium, Bulgaria, Canada, Croatia, Czech Republic, Denmark, Estonia, Finland, France, Germany, Greece, Hungary, Ireland, Italy, Japan, Latvia, Lithuania, Luxembourg, Malta, Netherlands, New Zealand, Norway , Poland, Portugal, Republic of Korea, Romania, Russian Federation, Slovakia, Slovenia, South Africa, Spain, Sweden, Switzerland, Turkey, Ukraine, United Kingdom and United States.

24. The Wassenaar Arrangement contributes to maintaining the technology lead by member states and denial/ delaying availability of technology to other states. The Wassenaar Arrangement has a detailed list of dual use technologies. The technologies have been classified in 99 categories in addition to a Sensitive and a Very Sensitive list. The nine categories are as follows:-

(a) Advanced materials.

(b) Material Processing.

(c) Electronics.

(d) Computers.

(e) Part 1 - Telecommunications.

Part 2 - Information Security.

(f) Sensors and LASERS.

(g) Navigation and Avionics.

(h) Marine.

(i) Aerospace and Propulsion.

The above lists are very detailed and run into 156 pages and are available at www.wassenaar.org/publicdouments.

25. In view of the Nuclear treaty with USA, efforts should be made to remove the embargo on import of dual use items and technologies to strengthen the industrial base.USA has relaxed its Export Control policies in ITAR (International Traffic in Arms Regulation) and Export Administration Regulations (EAR). The export controls in USA are not likely to be

diluted, in fact they might become stringent.The GAO office after a study of two years has recommended to the congress that the Executive & Legislative need to reexamine the system needed to help protect critical technologies export, post 9/11 in the long term security interest of the country (www.lasowskia@gao.gov).

26. To overcome the above problem it is import ant to conduct R&D to develop our own technology. Efforts in this direction have been made for the last 5 decade and a large number of R&D labs have been set up for developing defence related products. One of the major drawbacks has been the long product development cycle. When the product development cycle becomes too long, then it creates two major problems. Firstly the Product requirement in terms of user specification changes and we have to start again. Secondly the long cycle time increase the cost of project. Both the factors lead to import of equipment which are under development process in the country. This demoralizes the efforts of the indigenous R&D. India has set up fairly large Research and Development facilities with qualified and talented manpower. Some products have been developed through indigenous R&D efforts but 70 % of the strategic defence equipment s are imported. The projections for the procurement of defence equipment from abroad in the next decade are very high.

27. The offsets from our defence purchase in the next decade plus are likely to be of the order of 10 billion US Dollars. Policy for offsets was developed in 2006 and recently revised in Defence Procurement Policy-2008 (DPP-2008). The framework for strategies to realistically absorb the offsets to strengthen our Defence Industrial Base (DIB) requires a comprehensive study to aim at the long term objective of a sound Defence Industrial Base by 2025. The above issues have to be taken into account for developing the strategy to achieve the desired DIB by 2025?

28. The objective of the study is to answer the following main questions:-

(a) What is the present state of DIB in India?

(b) What should be the state of DIB by 2025?

(c) How is the gap to be bridged?

(d) How best can offsets be leveraged to develop the desired DIB?

(e) Who are the main players in public and private sectors with competencies to deliver results?

(f) How can the R & D base be developed and kept updated?

29. Major issues under consideration for the study are:-

- How to update the defence production system with indigenous manufacture of key products?
- Make the defence production competitive in terms of quality, cost and delivery.
- How to reduce the product development cycle?
- Technology acquisition, absorption and up gradation mechanism.
- Collaboration with private sector & foreign establishment.
- Technology forecasting.
- Likely gaps in technology and manufacturing capabilities keeping in view the requirements of the defence in future-2025.
- Manpower development keeping in view the future requirements.
- Building long term relationships with academia for basic R & D.

30. **Methodology.**

The subject of study is conceived and developed by CENJOWS. Further , the study has been conducted under the directions and supervision of CENJOWS. Expert s have been chosen by the leader of the study group. The experts have been chosen from the Defence services and private sector who have wide experience in the area ofTechnology management, R&D in the private, public sector and Defence .The expert s in turn have collect data from secondary as well as primary sources. Mainly open source material has been used for the study . The reports already submitted by dif ferent committees on the related issues were also consulted The study team expert s have collected information based on published literature and supported by discussion with various organizations in the private sector, public sector, Government departments and academic institutions .The team had the benefit of discussion with area expert s from Integrated Defence S taff. & large number of Ex. Defence personnel also. International dat a has been collected from the internet as well as the report s published by SIPRI and Rand Corporation. The international reports on the subject were studied on the internet also. Brain storming session was also organized to discuss the emerging technologies and the alternative means to accelerate the technology absorption process in the country.

31. **Limitations of the Study.**

Defence Industrial Base has many complex components which are governed by various organizations. It has wide geographical distribution in the country. It is spread over in the Government Departments, production and R &D units as well as the manpower involved in Science, Technology and its management. The technologies and equipment in the Defence Industrial Base are very wide. It was not possible to study all the sectors in detail. Therefore detailed study was restricted to few technologies only.

The concept of Offset policy is comparatively new in the country and it is in evolution stage. Though the policy was announce d first in the year 2006 but it was applied to few cases only. The policy has been further revised in Aug. 2008 and its ramifications are yet to be seen. The data on Indian experience in Offset utilization is negligible.

The Defence production was opened to the private sector in 2001. A large number of Letters of Intent and Industrial Licence have been issued but most of them are yet to be implemented. Therefore it is dificult to estimate the capacity of the private sector for Defence products.

The DIB covers many technologies used by the three services. Technology is changing at a fast rate and forecasting technologies has inherent uncertainties.

The R&D institutions project their success stories in great length but the data on failures is usually not available. This is applicable to private, public as well as Govt. R&D.

32. **Recommendations.**

The recommendations are as follows:-

I. India needs to reduce its Import dependence of military hardware from 70% to a minimum. It is projected that India will purchase approximately US $ 100 bn of military goods in the coming plan period and indigenous industry should help reduce a large portion of this outflow.

II. India has a well established infrastructure of defence production units which need to be vitalized so that they can absorb technology imported, this needs organizational reforms.

III. Technology denial regime will continue to manifest it self in one form or another , therefore self reliance in cutting edge technology has to be resorted to on priority. India has demonstrated its capability in the Nuclear and Computing technology as a result of denial and can do so again.

IV. Investment in technology development has to be in the top right quadrant of the technology development cycle. This will impart innovation advantage besides spin off advantages of leadership in technology and bargaining power . Commercial advantages are inbuilt in success in any area falling in this quadrant.

V. Policy changes will have to be made to implement drivers to growth.

VI. Multimode method of technology transfer may be adopted in critical areas.

Chapter 2

International Defence Technology Scenario

Introduction

1. Modernization and up gradation of technology is a continuous process to keep the armed forces to meet any challenges of tomorrow National security can only be achieved if Defence preparedness is total and self reliant, based on short technology development cycles. The existing defence industry is producing dated weaponry and in an inef ficient manner. It is beset with bureaucratic controls, no competition and poor HRD policies all in a priority sector. Private participation is virtually non existent. DIB consists of the complete array of strengths consisting of Design, Production, Maintenance, Logistics, SCM, Infrastructure, Training, Testing &evaluation, Outsourcing, Exports. The world is progressing towards a revolution in milit ary affairs (RMA) where precision weapons with stealth technology using sophisticated information and communication systems will be able to take real time decisions to t ake surgically precise operations. This is the challenge for Indian industry to meet in order to achieve unfettered national development.

2. While technology is at the root of self reliance it s efficient Management with sound HRD policies are the other two pillars without which this edifice cannot stand. Dependence on imports has other fallout such as, killing domestic industry, dependence on spares for maintenance, increasing further imports and most important discouraging R&D.

3. Discussing the Technological change and the future of warfare in the book by that name, Michael O'Hanlon st ates that in the 21st century the vision that most schools of thought accept are on the following premises:-

- Improvements in computers and electronics will make possible major advances in weapon and warfare, mostly in areas such as information processing and information networks but also in communication, Robotics, advanced munitions, and other technologies.
- Sensors will become radically more cap able, in ef fect making the battlefield

'Transparent'.

- Land vehicles, ships, rockets, and aircraft will become drastically lighter, fuel efficient, faster, and more stealthy, making combat forces far more deployable and lethal.
- New types of weaponry such as sp ace weapons, direct energy beams, and advanced biological agents will be developed and widely used.

From the above he draws two conclusions,

A. Properly exploited and integrated into military organisation tactics, and concepts of operations, these technical trends can soon add up to a revolution in military affairs.

B. Even less technologically sophisticated and wealthy countries will also benefit from the RMA and will acquire and learn to make good use of weapon of mass destruction, precision weapons, satellites, anti satellite, weapons, computer virus etc. they will exploit the big powers like USA or others who are dependent on big bases and other large assets by targeting them.

4. The route to acquire technology is by imports or developing indigenous capability. In the first case it is subject to exploitation by the seller or purchased at exorbitant cost since the purchase of military hardware does not t ake place under perfect-market conditions. Further imported technology is usually the second best and has political strings attached. If technology is transferred it cannot be absorbed without having a suitable base to do so. To obtain technology fast some countries obtain it by the ofset method. However Jurgen & Dunne in their research have found offset to be not dependable and most countries using this have not been benefited. Caution is therefore advocated. The answer is the second alternatives i.e. develop home industry.

5. Viewed from the macro level India has a large defence R&D base, large number of factories a civil R&D organisation, excellent talent and a very capable Private sector.......but very poorly managed with zero symbiotic approach.This is what bodes ill the Indian situation. This state, fortunately, is not beyond resolve. Over the period of years large organisation have developed having distinct culture, influences and territories. Cartels and power centers have developed that will prove invincible.Their existence has to be accepted. Therefore to achieve anything in this situation the start, necessarily, has to be at the macro level.

6. Further the DIB has to be examined keeping in view the International scenario. For the purpose of bench marking the study has focused on USA and China, though we have observed the development in other leading countries such as Russia and Korea .USA

has been chosen since it is the world leader in Defence Production, technology as well its export. China is another neighboring country which is growing very fast in this area and impacts India directly. Brazil is another country in a situation similar to India, that has successfully developed its aircraft industry by transfer of technology.

Drivers and Barriers for Technology Development and its Application

7. GlobalTechnology Revolution has been examined in detail by National Security Research Division of Rand Corporation in in2006.They examined 56 technologies which were possible by 2020 and out of these they identified 16 technologies that have the greatest combined likelihood of being developed. These technologies would have effect on multiple sectors like water, food, land, population, governance, social structure, health, economic, development, education, defence and conflict, environment and pollution. These key technologies are as follows:-

- Cheap solar energy.
- Rural wireless communication.
- Communication devices for ubiquitous information access.
- Genetically modified crops.
- Rapid bioassays.
- Filters and catalysts.
- Targeted drug delivery.
- Cheap autonomous housing.
- Green manufacturing.
- Ubiquitous radio frequency identification (RFID) tagging of commercial products and individuals.
- Hybrid vehicles.
- Pervasive sensors.
- Tissue engineering.
- Improved diagnostic and surgical methods.
- Wearable computers.
- Quantum cryptography.

8. The study has examined the technical feasibility as well as implementation feasibility of the above technologies across the 29 representative countries of the world. Technical

feasibility is defined as the likelihood of that the application will be possible on commercial basis by 2020. Implementation feasibility is the net of all non-technical barriers and enablers, such as market demand, cost. infrastructure, policies and regulations. Represent ative countries across the regions of the world were selected for analysis as follows:-

- Asia (China, India, Indonesia, Japan, South Korea, Nepal, Pakistan).
- Oceania (Australia, Fiji).
- North Africa, and the Middle East -(Egypt, Iran, Israel, Jordan).
- Europe- (Georgia, Germany, Poland, Russia, Turkey.
- Africa- (Cameroon, Chad, Kenya, South Africa).
- North America -(Canada, Mexico, United States).
- Central and South America and the Caribbean -(Brazil, Chile, Colombia, Dominican Republic).

9. The ten major driver and barriers that the countries may face through the 2020 have been identified as follows:-

- Cost and Financing.
- Laws and Policies.
- Social value, public opinion, and politics.
- Infrastructure.
- Privacy concerns.
- Use of resources and environmental health.
- R&D investment.
- Education and literacy.
- Population and Demographics.
- Governance and Political Stability.

10. The conclusions of the study are that the capacity to implement the key technologies would depend upon technical feasibility as well as the drivers and barriers faced by a country. The likely scenario projected by the study by 2020 is as follows. Countries which have high S&T capacity and lesser barriers may be able to implement all the 16 key

technologies. These would include:-

- USA.
- Canada.
- Germany.
- Japan.
- South Korea.
- Australia.
- Israel.

11. The countries which would have high S&Tcapability but the number of barriers would exceed the drivers, making it difficult to implement all the key technologies. They may be able to implement 12 out of the 16 key technologies. These would include China, India, Poland & Russia.

12. The countries which will lack high S&T capability, few barriers and few drivers. They may be able to implement only 7 out of the 16 key technologies. These would include Brazil, Chile, Mexico, Turkey, South Africa, Indonesia & Columbia.

13. The countries lagging in S andT and having too many barriers and few drivers may be able to implement only 5 out of the 16 key technologies. These would include Cameroon, Chad, Kenya, Dominican Republic, Georgia, Fiji, Egypt, Iran, Jordan, Nep al & Pakistan. The selected countries 'cap acity to implement the top 16 technology applications are summed up in the diagram given below.

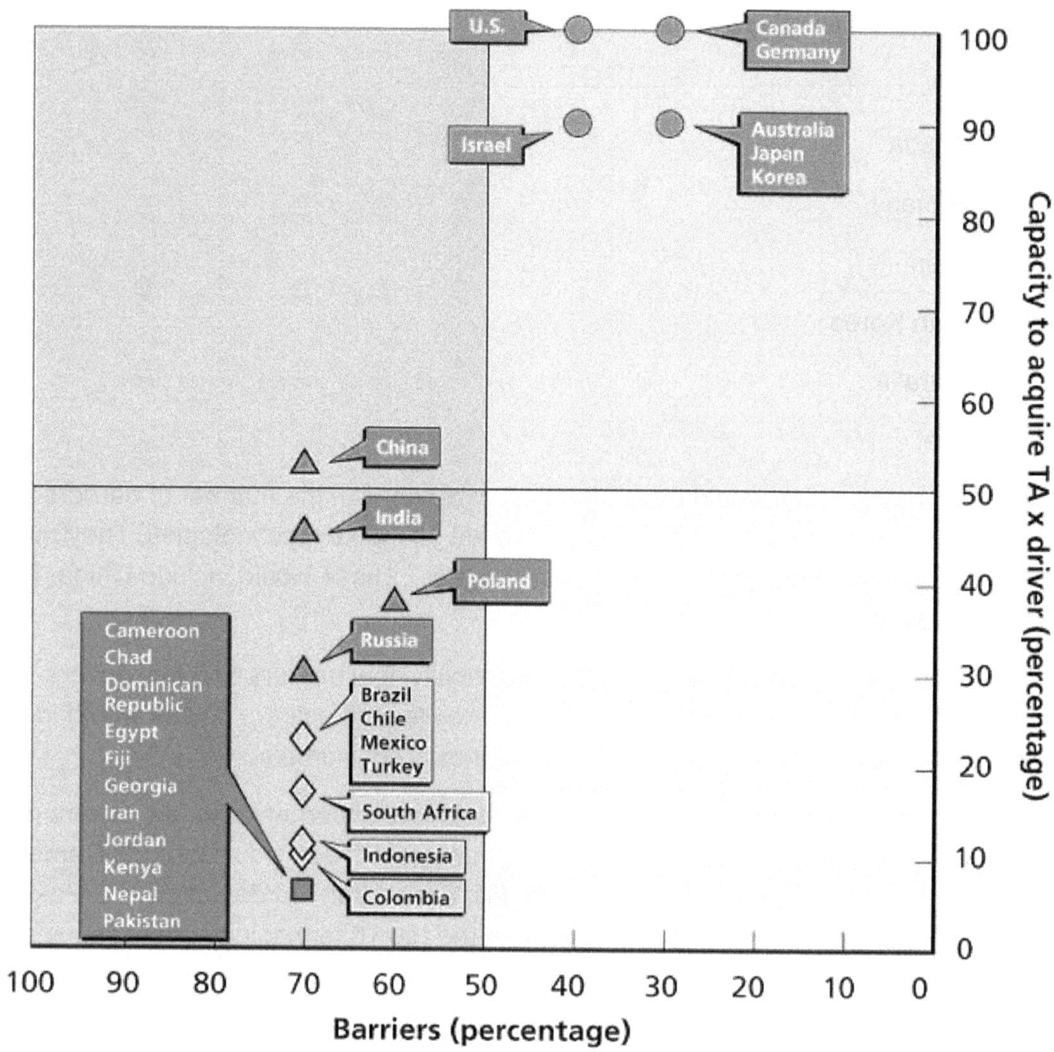

14. Out of the four quadrants above the one with minimum barrier and minimum capacity and the second with highest barrier and high capacity are theoretical and may not exist unless external strong forces compel it. Therefore most countries fall into those with low capacity due to high barriers and high capacity due to low barriers. Therefore capacity to have a higher or lower capacity to acquire technology has an inverse correlation with existing barriers.

15. India falls in the category of low capacity just below China and ahead of Russia,

Poland, Brazil, Indonesia etc. the lowest are Pakistan, Nepal, Kenya Egypt Cameron etc. Out of the ten barriers, identified, that hinder technology growth the single barrier faced in the Indian context is 'Laws and Policies'. While these exist on paper and as amongst the best in the world implementation is exactly the opposite.The stranglehold of the bureaucracy and long delays in court s of law virtually cause them to be non-existent and result in extraneous factors calling the shots. Reform of this barrier should enable India to reach in the fourth quadrant along with Israel, Australia, Japan and Korea, just below the top three i.e USA, Canada and Germany.

16. China developed post liberalisation to become the fastest growing economy. Indian economy demonstrated its capability within ten years post liberalization resulting in high growth of industry. There is no reason why governmental shackles should not be loosened and same opportunity afforded to the defence industry.

17. Science and technology is accorded high import ance by US defence dep artment. Military research & development is an essential process identified with the revolution in military affairs (RMA) which is evolving towards Network-centric solutions.Today there is a move to shift emphasis from the traditional R&D of defined weapon systems to a greater military exploitation of science and technology also referred as science & technology base military innovation(S&T). A common term in use is military research and technology (R&T) as a complement to R&D. the T in R&T emphasis the development of military technology and compared to 'exploratory development' illustrated by technical demonstrators in contrast to traditional milit ary prototypes. Countries in Europe st arted to formulate milit ary R&T strategies in 1900's and NATO R&T organization was formed in 1998. S&T based military innovation implies the military exploitation of the indigenous and imported research. The S&T effort have been found to be relatively cheaper since it adds to the ongoing research. Amongst other reasons this shif t was due to change of doctrines from threat based to capability based strategies. This also factors in new rules of engagement including asymmetric threats and terrorism. Another reason for this shift to R&T is due to pressure on reducing milit ary expenditure. The difference between milit ary and civil research is small and common high technology civil end use items are being employed for milit ary purpose.

18. It therefore implies an increasing involvement of non military p articipants in military innovation such as universities, private, Public and commercial organizations. The USA has more than twenty organizations devoted to research in multi discipline military related fields to basic and applied research. It also has of fices abroad to cooperate with other countries on emerging ideas on technology In March 2003 the defence department set up

the Technology planning Guidance (TPG) a budget tool to establish priority areas for S&T. It also set up the USArmy Science officer position for liaison to bring quickly the operational experience to the research labs and help in implementing new technologies.The potential of contribution by the private sector especially by the SME's must not be underestimated in the Indian context.

19. The importance of R&T for the military program is reflected in the budget allocations for Research Development, Testing and Evaluation (RDT&E) has increased 76 % since 2001 as below:-

Military Spending on RDT&E by USA from 2001 to 2007 in $ bn.

Year	2001	2002	2003	2004	2005	2006	2007	2008
Amount	40.5	44.4	53.1	60.8	65.7	68.6	71.1	69.6

Source: Bojrn Hagelin in SIPRI 2008

Brief on the international scenario is discussed below.

Defence Technology Scenario-USA

20. Introduction- At present the Defence Industrial Base in USA is the responsibility of Bureau of Industry and Security (BIS) within the Department of Commerce.BIS is the focal point for developing, promoting, and implementing policies that ensure a strong, technologically superior US Industrial Base. The key responsibilities of BIS include the following component:-

- Defence Priorities and Allocation System Program (DPASS).
- This is for timely availability of industrial resources to meet current national defence and emergency preparedness. DPASS also provides an operating system to support rapid industrial response in a national emergency.
- Defence Trade Advocacy to support US companies competing in the global defence market.

- Best Manufacturing Practices Programme to promote the best practices used in industry, government and academia.
- Defence Industrial Capability and Technology Assessment. BIS conducts industry analysis to assess cap abilities of the US industrial base to support the national defence. Industry specific surveys are used to provide essential dat a for these comprehensive assessments. Typical ongoing studies are in the area of 5-axis machine tools, counterfeit electronics, and critical microchip. In the last 10 years the area of studies have been space industry, aromatic polyamides, imaging and sensor industry, cartridge and propellant actuated devices, Air Force C-17 PROGRAM, munitions power source (batteries),Ariel delivery parachutes, theatre support vessel procurement, US textile and app arel industry, Biotechnology assessment, assistive technologies, welding in manufacturing and mining industry Ball and roller bearing industry , technology transfer to China, Opt-electronics, emergency aircraft ejection seats, semiconductor processing materials, Encryption, superconductivity, artificial intelligence etc.
- Monitoring the impact of Offsets in Defence Trade Annual Reports.
- Business and technology enhancement opportunities. BIS takes initiative to matches industry with government agencies and leading National Labs to access technology transfer, manufacturing enhancement and exporting resources.The goal is to make the US industry competitive in the global market.

21. The dominance of theAmerican defence industry is a result of the economy based on defence industry post world war II for almost four decades.This was due to cold war threat in Europe. The aircraft industry was the most dominating, and further diversified into commercial airliners, S pace related product s and other high technology product s and services that required heavy investments in R&D. Therefore large companies which form the backbone of the Defence Industrial Base of USA are in the private sector . In fact majority of of fset programs in other countries are sponsored by the aircraf t industry of USA. The SIPRI list of top 100 arms producing companies is dominated by with 41 US companies accounting for 63 % of the top 100's sale.

Operational Capabilities

22. Four operational cap abilities and their enabling technologies that can support the range of future milit ary missions have been identified by Rand corporation as follows. (www.rand.org)

1. To understand how individuals, groups, societies and nations behave and then use this information to improve performance of US forces. Key enablers include automated language processing and human, social, cultural and behavior modeling.

2. Enhance capabilities to observe people, things, and activities in urban and other tough terrains and to record and recall the data. This will reduce sanctuaries where adversaries hide and draw support for their operations, and reduce their ability as irregular forces to "hide in plain sight." For this, needed are close-in sensors and the soldier on the ground empowered to be a powerful collector.

3. Extracting actionable information hidden in massive data much more rapidly than is done today. This capability would be critically important to commanders at all levels, intelligence analysis and soldiers and marines on patrol.

4. Producing effects-offensive and defensive, kinetic and non-kinetic, lethal and non-lethal-tailored rapidly to circumstances in order to achieve the desired and avoid counterproductive.

23. The focus of USA is on a set of four critical capabilities for 21st century.

Following technology areas are important to acquire the four capabilities:-

Critical Capability	Technology Area
Human Terrain Preparation	*Rapid Training and continuous learning *Automated Language Processing *Human, Social, Cultural and Behavior Modeling
Ubiquitous Observation and Recording	* Day Night All Weather wide Area Surveillance * Close in Sensors and Tagging Systems * Soldier as a Collector
Contextual Exploitation	* Mega Scale Data Management * Situation dependent Information Extraction * Human/System Collaboration
Rapidly Tailored Effects	* Influence Operations
	* Time-critical conventional strike
	* WMD Protection and Mitigation

24. **Following are the Three key Technology Areas:-**

- Day and night all-weather wide area surveillance, particularly in urban areas and foliage.
- Close-in sensors and tagging systems.
- Soldiers-as-collectors within a network.

25. **Contextual Exploitation.** A new set of information-rich exploitation and collaboration tools are needed to bridge the functional capabilities of ubiquitous observation and rapidly tailored delivery of scaled effects. Three enabling technologies are:-

- Mega-scale data management.
- Situation dependent information extraction.
- Human / system collaboration.

26. **Rapidly Tailored Effects.** The complexities of the 21st century and the challenges in countering a wide range of adversarys, demands an extensive tool kit of action options. They are as follows:-

- Enhancing the ability to conduct non-kinetic operations aimed at influencing the local populace.
- Delivery of conventional strikes with great precision and timeliness from afar.
- Mitigating the effects of the Weapons of Mass Destruction (WMD) attacks.

27. **Mission-Focused Portfolios.** For linking future requirements to capabilities the following is required.

- Anticipate the technological possibilities across the domains, time horizons, and challenges for the country, and its potential adversaries.
- Identify those technological possibilities that could create new key capabilities important to meeting broad challenges.
- Identify disconnects between desired capabilities and technology availability.
- Propose investment strategies to secure the technologies needed over time.
- Shape decisions for disinvestment driven by outdated concepts.

28. **Mission Oriented Approach to Preparing Strategic Technology Plan:** A team should be assigned the following mission:-

- Establish the state of current technology base.
- Derive the projected technology needs identified overtime.
- Enumerate the resulting technology gaps and the resources (time, money people) and partnership with others, necessary to close these gaps and create opportunities.
- Identify metrics necessary to track closure of gaps.

Implied in this process is the need to anticipate technology advances and to anticipate how potential adversaries will adopt to these advances.This task is dynamic, as is the task of tracking metrics and continually updating the plan.

Strategic technology plan should cover a period of over 20 years across all disciplines. New technology thrusts should be identified, which will enable new capabilities. The Indian Military leadership is seized of these imperatives in the developing military-strategic milieu and factors them into their LTIPP but implementation is hindered more due to integrated policy & planning than lack of any resource.

Important Technologies have been identified in various studies done in USA. Consolidated list of the technologies is given below.

29. High altitude long endurance platforms, Giga-pixel optical imaging, space based GMTI/ SAR, Active/passive hyper spectral sensors, Foliage penetration sensors, Stealthy precision delivery platforms, Micro sensor technologies, High performance high efficiency signal processing, Ultra high density packaging for millimeter scale sensors and tags, Native signature recognition at long ranges (human and object), Efficient energy storing technology, Miniature sensor and tag technology, High-density packaging for centimeter-scale devices, Interactive automated debriefing,All-domain precision geolocation, Soldier centric communication and networking technology Body borne flexible displays, Multilevel security and accreditation, Human guided algorithms, Contextual analysis and intent recognition; Entity-relationship and pattern analysis, Data-to-information-to-target recognition, Data management and fusion from very diverse sensors, Information retrieval and knowledge discovery, Natural man-machine interface, Knowledge representation, Kinetic and non-kinetic cause/efect models, Decision support tools incorporating complexity and ambiguity, Storytelling-gisting and advanced visualization, Gaming technology for campaign planning and targeting / shaping, Directed energy weapons, Hypersonic delivery vehicles, Ballistic missile technology, Rapid diagnostics and environmental monitoring

including standoff biological and radiation detection, Broad spectrum medical countermeasures, Decontamination technologies, Nuclear weapons affects prediction-mitigation.

DARPA Model

30. The Defense Advanced Research Projects Agency (DARPA) is an agency of the United States Department of Defense responsible for the development of new technology for use by the military (www.darpa.mil). DARPA has been responsible for funding the development of many technologies which have had a major impact on the world, including computer networking, as well as NLS, which was both the first hypertext system, and an important precursor to the contemporary ubiquitous graphical user interface.

31. Its original name was simply *Advanced Research Projects Agency (ARPA),* but it was renamed DARPA (for Defense) on March 23, 1972, then back to ARPA on February 22, 1993, and then back to DARPA again on March 11, 1996.

32. DARPA's original mission, established in 1958, was to prevent technological surprise like the launch of Sputnik, which signaled that the Soviets had beaten the U.S. into space. The mission statement has evolved over time. Today, DARPA's mission is still to prevent technological surprise to the US, but also to create technological surprise for their enemies.

33. DARPA is independent from other more conventional military R&D and reports directly to senior Department of Defense management. DARPA has around 240 personnel (about 140 technical) directly managing a $3.2 billion budget. These figures are "on average" since DARPA focuses on short-term (two to four-year) projects run by small, purpose-built teams.

DARPA's Mission

34. DARPA is a Defense Agency with a unique role within DoD. DARPA is not tied to a specific operational mission: DARPA supplies technological options for the entire Department, and is designed to be the "technological engine" for transforming DoD. DoD needs a place like DARPA whose *only* charter is radical innovation. DARPA looks beyond today's known needs and requirements.

35. As military historian John Chambers noted, "None of the most important weapons transforming warfare in the 20th century – the airplane, tank, radar, jet engine, helicopter, electronic computer, not even the atomic bomb – owed its initial development to a doctrinal requirement or request of the military.". And to this list, DARPA would add unmanned systems, Global Positioning System (GPS) and Internet technologies. DARPA's approach

is to imagine what capabilities a military commander might want in the future and accelerate those capabilities into being through technology demonstrations.

36. The DARPA's key characteristics are as follows: -

- Small and flexible: DARPA has only about 140 technical professionals.
- Flat organization: DARP A avoids hierarchy , essentially operating at only two management levels to ensure the free and rapid flow of information and ideas, and rapid decision-making.
- Autonomy and freedom from bureaucratic impediments: DARPA has an exemption from Title V civilian personnel specifications, which provides for a direct hiring authority to hire talent with the expediency not allowed by the standard civil service process.
- Eclectic, world-class technical staff and performers: DARPA seeks great talent and ideas from industry, universities, government laboratories, and individuals, mixing disciplines and theoretical and experimental strengths. DARP A neither owns nor operates any laboratories or facilities, and the overwhelming majority of the research it sponsors is done in industry and universities. Very little of DARPA's research is performed at government labs.
- Teams and networks: At its very best, DARPA creates and sustains great teams of researchers from dif ferent disciplines that collaborate and share in the teams' advances.
- Hiring continuity and change: DARPA's technical staff is hired or assigned for four to six years. Like any strong organization, DARPA mixes experience and change. It retains a base of experienced experts – its Office Directors and support staff – who are knowledgeable about DOD.
- Project-based assignments organized around a challenge model: Although individual projects typically last three to five years, major technological challenges may be addressed over longer time periods, ensuring patient investment on a series of focused steps and keeping teams together for ongoing collaboration. Continued funding for DARPA projects is based on p assing specific milestones, sometimes called "go/no-go." They have the provision for closing projects.
- Outsourced support personnel: DARPA extensively leverages technical, contracting, and administrative services from other DoD agencies and branches of the military.

This provides DARPA the flexibility to get into and out of an area without the burden of sustaining staff,

- Outstanding program managers: The best DARPA program managers have always been freewheeling zealots in pursuit of their goals. The Director's most important task is to recruit and hire very creative people with big ideas, and empower them.
- Acceptance of failure: DARP A pursues breakthrough opportunities and is very tolerant of technical failure if the payoff from success will be great enough.
- Orientation to revolutionary breakthroughs in a connected approach: DARP A historically has focused not on increment al but radical innovation. It emphasizes high-risk investment, moves from fundament al technological advances to prototyping, and then hands of f the system development and production to the military services or the commercial sector.
- Mix of connected collaborators: DARPA typically builds strong teams and networks of collaborators, bringing in a range of technical expertise and applicable disciplines, and involving university researchers and technology firms that are of ten not significant defense contractors or beltway consultants.

Current Program Offices

37. DARPA has five program offices, all of which report to the DARPA director.

- **The Defense Sciences Office (DSO).** V igorously pursues the most promising technologies within a broad spectrum of the science and engineering research communities and develops those technologies into important, radically new military capabilities.
- **The Information Processing Techniques Office (IPTO).** Focuses on inventing the networking, computing, and sofware technologies vital to ensuring DOD military superiority.
- **The Microsystems T echnology Office (MT O).** Mission focuses on the heterogeneous microchip-scale integration of electronics, photonics, and micro electro mechanical systems (MEMS).Their high risk/high payof technology is aimed at solving the national level problems of protection from biological, chemical and information att ack and to provide operational dominance for mobile distributed command and control, combined manned/unmanned warfare, and dynamic, adaptive military planning and execution.

- **The Strategic Technology Office (STO)** - In July 2006, the SPO and ATO were merged into a single Strategic Technology Office (STO) that complements the Tactical Technology Office (TTO) as one of the two "systems" offices.
- **The Tactical Technology Office (TTO)** - engages in high-risk, high-payoff advanced military research, emphasizing the "system" and "subsystem" approach to the development of aeronautic, space, and land systems as well as embedded processors and control systems. This research includes an effort within the TTO to develop a small satellite *launch vehicle*. This vehicle is under development by *Air Launch LLC*. (*Source-Wikipedia & DARPA website www.darpa.mil*)

DARPA could be an excellent model to encourage developing technologies of the future.

Relevance of US Technologies to India

38. Some of the important technologies listed above are relevant for the armed forces in India today. The rest of them are expected to be relevant in the next 5 to 10 years. This view is further reinforced by views of several defence analysts during the Seminar in Mar 2004 ('Army 2020' - Centre for Land Warfare Studies, Delhi - 2005).

Defence Technology Scenario-CHINA

39. In China the World-wide Revolution in Military Affairs (RMA) is gaining momentum. The forms of war are undergoing charges from mechanization to informalization. Informalization has become the key factor in enhancing the war fighting capabilities of the armed forces. Going with the tide of the worlds' military development, the Peoples Liberation

---

References

1. Defense Technology scenario-USA" described above is derived from "21 st Century Strategic Technology Vectors"-February published in USA 2007,

2. Defense Industrial Base, Homeland Security, Department of Defense, May 2007

3. The U.S defense Industrial Base-National Security Implications of a Global World, June 2, 2005 Washington D.C

4. Evolution of the US Defense Industry, Michael D.Rich-Oct1990

5. Army 2020 – Centre for land warfare studies - Delhi.

6. www.rand.org

7. www.darpa.mil

Army (PLA) shall gradually achieve transition from mechanization to informalization.They plan to build a strong military by means of science and technology, by transforming it from manpower intensive to technology intensive one. China conducts explorations of frontier technology and future oriented research, so as to exp and its technological reserve. It employs new and high technologies to remold milit ary industrial enterprises. It aims at shortening the development cycle and reduces the cost of products. It gives priority to the development of dual purpose high-tech industries and promotes and promotes two – way transfer of milit ary and high technology . China encourages co-operation with large multinational companies, and raise the technological of the defence – related science, technology and industry. (www.cfr.org/publication/15577/evolution)

40. China' s "grand strategy" for improving defence-industrial cap abilities is based on selective modernization to focus on making breakthroughs in certain key areas of weapons, capabilities like: aerospace, missiles, electronic technology, Command, Control, Computers, Communications, Intelligence, Surveillance, and Reconnaissance. (wwwguardian.co.uk/ technology/2008)

41. They have done major progress in the field of ship building, aviation (excluding turbojet engines), Missiles, space satellites, UAVs and Cyber warfare.

42. New defence-industrial paradigm - The Chinese military is in the midst of C4I (Command, Control, Communications, Computers and Intelligence) revolution, characterized by the whole-sale shift to digital, secure communications via fiber-optic cable, satellite, microwave and encrypted high frequency radio (www .technology.timesonline.co.uk). Information technology is now universally recognized as the core of future warfare. Changes which are taking place are as follows:-

FROM	**TO**
1. Analogue Technology	1. Digital Technology
2. Electric Cables	2. Fiber-optic Cable
3. Mechano-electrical switches	3. Programme-Controlled switches
4. Single-function terminals	4. Multi- function terminals
5. Single-tasking networks	5. Multi-tasking networks
6. Manual operations	6.Automated and intelligent network management

43. The military information network envisions the fusion of traditional telecommunication networks with advanced computer networks. China's I.T. industry is successfully coming up to meet the needs of military in this area. Following points, among others, are important for "DIB-2025 for India", based on Chinese experience:-

- China's long term goal is "self reliance in defence production"
- China's defence industry now has the potential to become more competitive with defence industries of the world's advanced military powers.
- Their aim is to inject into their DIB, the principles of "competition, supervision, evaluation and encouragement"
- They are expanding partnerships with MNC's, civilian universities and research institutes to improve educational training relevant to military R & D.
- They have reorganized defence industries on "Holding Company" pattern with these companies having production and R & D facilities under their control. Hence, one authority is responsible for bringing out new products. This system appears to be more efficient than the one being followed in India.

Russian Defence Scenario

44. Russia / erstwhile USSR have been the traditional source for defence equipment and technology to India. They have a large infrastructure for production as well as R&D for Defence industry. Brief on the Russian defence scenario is given in this chapter. Prior to the recent wave of restructuring, Russian defence industry could be placed in three broad categories viz. research institutions, design organizations and production facilities.

- Research institutions would focus on basic and applied scientific research.
- Design organizations included design bureaus and technological institutes. Design bureaus planned new products and machines, although some conducted research. Technological institutes had responsibility for designing new processes, installations and machinery. The research organizations are the heart of Russian military research and development. They take new weapons and military material projects from concept to prototype, and then hand them off to the production enterprises. Production enterprises do prototype construction, production runs and modifications.
- Production facilities manufactured the new products or applied the process developed by the research and design facilities.

45. Output and testing of industrial prototypes, industrial innovation processes or small batch production prior to the stage of mass production occurred in experimental production or pilot plants. These functioned independently or were attached to production facilities, research institutions or design organization.

46. In 1987, an estimated 450 R & D organizations were working exclusively on military projects, like: generation of radio frequency energy, laser Programme and particle beam research, all applicable to future battlefield weapons. In addition, about fifty major design bureaus and thousands of plants were making military items exclusively. With the end of cold war, Russian defence plants lost nearly 80 percent of the Government funding. At the same time, they lost export earnings due to loss of traditional soviet block markets. The Russian defence industry consisted of over 1,700 firms in 1997 and employed some 2.5 million workers. The Russian defence industry underwent changes which included a shift from production of military to commercial goods; the restructuring of firms into a mixture of holding companies, joint ventures; and a shift towards integration into the world economy

47. Since the early 1990's, one of the principal thrust areas of restructuring has been the formation of new corporate structures that would unite design and manufacturing facilities into financial industrial groups. The intent was to integrate corporations on the pattern of western companies, or the South Korean Chaebols, to replace the soviet centralized defence industrial structure. These new entities would span multiple enterprises, though on a scale smaller than the entire industrial sector encompassed by the soviet era ministries.

48. The Russian government at one time planned to privatize three – fourths of its defence enterprises by 1994 and following four categories were created:-

- Fully privatized facilities that no longer engaged in military work.
- Fully privatized enterprises that were still engaged in military work.
- Joint-stock companies in which the state held a stake.
- Fully state-owned enterprises that remained directly subordinate to state.

49. By 1995, only 21 percent of the enterprises in the aviation industry were fully state owned; the armaments industry had 35 percent state ownership; in the shipbuilding industry it was 40 percent; it was 54 percent in the missile – space industry, and 82 percent in the munitions and special chemicals industry. A government decree of 13 July 1997 approved a total of 480 industries that would not be subject to privatization. Over 450 enterprises, mainly research institutes and design bureaus, were to remain government owned. Of

these, 45 were in aviation industry 60 in missile – space industry, 54 in shipbuilding industry and 168 in the communications and electronic industry.

50. Despite the general crisis besetting the defence industry examples of highly advanced military technology continued to emerge from Russia's defence plants in mid-1990. The T-90 main battle tank, the most modern tank in the army arsenal, went into production in 1993. It features a new generation of amour on its hull and turret.

51. Russia's submarine technology developed faster in mid-1990's than the western experts had expected. It is believed that more than half of the 1996 numbers (around 99 nos.) was capable of moving undetected into western sea-lanes. The new Severodvinsk class is the-state-of-the art submarine that allegedly is so quiet that it eliminates the US technical lead in this area, and it is armed with the new 650 mm Shkval rocket that travels at 200 knots underwater. A project Vasen – class multipurpose attack nuclear – powered submarine is preparing to hit water . It is fourth generation and will cause revolution in submarine shipbuilding. Russia has a strategy for shipbuilding until 2030, under which warship production is to increase by 50%. In February 2006, the frigate Admiral Sergei Gorshkov had its keel laid down. It is a surface ship intended for long-range operations in distant areas. The navy has plans to acquire 20 such ships.

52. As a strategy, Russia wants to drop the use of specialized ships capable of fighting only submarines or aircraft carriers and go over to multipurpose fighting units meant to carry out wide range of missions away from home. Such ships will be assembled from modular units, and their weapons and equipment will be unified for all types of combat craft. This will not only facilitate the provision of spare parts and ammunition, but also maintenance, repairs and modernization. They plan to build six aircraft carriers, which would make Russian Navy, the second in the world in combat capability. Three nuclear submarines of the fourth generation are currently under construction. They will have much better armaments and equipment.

53. R & D programmes of Russia include work on highly accurate precision – guided munitions, MIG-MAPO 1.42 single – seat multi role stealth fighter air-craft, new ICBM's etc. As mentioned earlier, Russia has been restructuring its defence industries since early 1990's. In 2007, further steps were taken to consolidate the aircraft and shipbuilding industries.

54. The United Aircraft Corporation (UAC), which consolidates most of Russia's Civil, and military fixed-wing aircraft design and production assets began its operation in 2007. The company which is majority-owned by Russian state will include 20 companies. The largest

component of UAC is "Sukhoi".The United Shipbuilding Corporation (USC) was established in 2007 as an entirely state – owned company. It amalgamates all wholly or partially state-owned shipyards and design bureaus for surface ships. Decision making responsibility, which is currently dispersed through several agencies in this sector, will be centralized. A third state – owned conglomerate "Rostekhnologic", was established in 2007. It combines the state arms export company with several military and civilian production operations, air defence and military electronics company, engineering equipment company, titanium producer etc. The new structure aims at boosting R&D and coordinating arms export activities with industrial supply chain. To illustrate the consolidation process further, the operations of two more companies, "The Tactical Missile Corporation" and "National Innovation Company New Energy Projects", have been briefly explained in the following pages.

55. The Tactical Missiles Corporation (TMC) is the leading Russian association of companies for the development and manufacturing of guided missiles, as well as, ship and shore – based tactical missile systems. The corporation was established in line with the national programme for the development of the defence – Industrial complex 2002 – 2006, which encouraged the formation of several large integrated structures in the sector. The corporation consists of 19 companies, including those involved in the final stages of production, like the TMC flagship enterprise (formerly called Zvezda – strela), Vympel, Raduga, Region, and KB Mash. The flagship enterprise specializes in the development and manufacturing of air-to-ground and ship-to-ship missiles, as well as, the creation of ship-borne missile systems. Vympel is the leading manufacturer of air-to-air missiles. Raduga is one of the oldest Russian enterprises for making air-to-ground missiles for long-range aviation and frontline fighters, as well as, deck – launched anti – ship missiles, Region makes guided bombs, antiship and antisubmarine missiles, and antisubmarine defence systems. Region also makes a unique antiship missile system based on high – speed underwater Shkval – E missile family. KBMash develops shore – based, mobile missile systems. Main clients of TMC are those countries that import Russia fighters of SU and MiG family, like China, India, Algeria and Venezuela.

56. TMC employed 21,249 persons in 2007, with income of 28,402 million rubles. The corporation is engaged in the entire life cycle of its products, from the development of design documentation, the making of prototypes, all the tests, series production, repair, and modernization, all the way through to its scrapping. It is currently implementing a comprehensive programme for the development of advanced aviation armament, including weapons for the next generation fighter. Another programme for the developments of advanced naval weaponry for the period 2010 – 2015 is under development. TMC is an

example of the successful reform of the Russian defence – industrial complex leading towards the creation of large integrated structures. S table export earnings together with growing domestic procurement for the armed forces allowTMC to look into the future with confidence.

57. On the whole, Russian defence industry has managed transition from "Cost no object" of soviet era to global market at extra – ordinary pace.

National Innovation Company New Energy Project (NIC NEP)

58. Over 20 enterprises of the Russian defence industry complex are p articipating in implementation of the hydrogen power engineering programme. National defence industry has experience in the field of hydrogen power engineering. The prototype of TU – 155 aircraft was equipped with hydrogen engines. Plans were on for development of a submarine with an anaerobic air – independent power plant consuming hydrogen.

59. NICNEP is an investment and operating company that was created to implement the investment programme "Hydrogen Power and fuel cells". Principal activities of the company are: R & D in hydrogen power and fuel cells; formulation of project s in this field; market research for domestic and international demand; and organization of manufacturing and sales. It is developing technologies in co – operation with over 50 scientific organizations of the Russian Academy of sciences and industrial design bodies for the implement ation of these tasks.

60. Considering the course of restructuring which Russian defence industry is following, it is worth noting that they are forming holding companies which have both production and design units under their control, and in one case export too. India too could consider adopting this pattern for better coordination among various wings, which may cut down development time of defence systems. We too could give up the old Russian model; we still follow even though Russia has given it up. For many years now India had to import defence equipment,

References:

- Reforms in Russian Defence Industry – Baidya Bikash Basu, Researcher IDSA, updated & maintained by webmaster, August 24, 2000.
- "Russia's Navy gets ambitions" by Nikita Petrov, published August 2007 – India strategic.
- "MOSCOW Defence Brief" by Sergey Denisentser
- SIPRI Yearbook 2008
- A Profile of the Tactical Missile Corporation – by Sergey Denisentsev.

even though largely it was within India's capability to design and produce such equipment. Delay in commercializing the products had been one of the important reasons. Formation of holding companies could perhaps eliminate one of the reasons for delay.

61. Summary of the Learnings from Russia are as follows:-

- Since early 1980's, one of the principal thrust areas of restructuring has been the formation of new corporate structures that unite design and manufacturing facilities into financial industrial groups to replace the Soviet centralized defence industrial structure.
- They have set up new entities that span multiple enterprises, hence several large integrated structures have been formed
- The Russian industry has managed the transition from 'Cost no object 'of Soviet era to global market at extraordinary pace
- They have also privatized some of the defence enterprises to make them competitive.
- India has continued with its model of DIB taken from the Soviet era, whereas the Russia has given it up.They have now specialized organizations in different sectors which have production, R&D and marketing etc under over all control of the Government.
- India has DRDO, CSIR, DST, Ordnance factories, PSU's-all run as independent entities. Results are there to see. In spite of large manpower and reasonable infrastructure, we still import 70 % of our defence needs and our defence products are at low level of technology and not competitive globally.
- India should seriously consider adopting this pattern for better coordination among the various wings, which may cut down development time and cost of defence systems. This would cut down the imports to achieve self reliance.
- They are forming holding companies which have both production and design units under their control and in some cases export units too
- In India too we have started permitting but not encouraging industries for defence in the private sector. The pace in the growth of private sector is too slow. The private sector should be allowed to offer competition to public sector enterprises. This may not happen at the current rate of development.There is a need to create a responsive environment for the private sector to grow. It may be noted that the

fields in which private sector was encouraged, the progress has been significant as in the case of IT, automobile and pharmaceuticals sectors.

Recommendations

62. The recommendations are as follows:-

(a) Reform of the bureaucracy and good governance is the biggest hurdle to develop DIB in India. India can come to the level of top six countries with these reforms.

(b) Private Industry must play a greater role in defence industry and the situation from '*granting permission*' needs to be enhanced to ' *encouraging growth*' must be adopted.

(c) There is a need to establish an office on lines of 'Bureau of Industry and Security' in the ministry of defence coordinated by HQ IDS to coordinate Inter-agency and industry developments affecting defence capability.

(d) Merging requirements of Operational capability and technology are known to defence HQ's and factored in the L TIPP these need to be implemented with Policy and Budgetary support between MoD, HQ IDS, MoF and Planning commission.

(e) DARPA has been very successful in developing defence technology and there is a need to examine the feasibility of adopting a similar model with modifications in India.

(f) China has made a political commitment of a 'Grand strategy ' to modernize the defence forces, a political and national commitment needs to be made in India by way of enunciating a National security policy and policy on DIB growth.

(g) The leading countries ie USA, Russia and China have or are converting existing Public defence production units into Corporations or holding companies successfully. India needs to follow suit under the national policy of dis-investment of the PSU's by according priority to defence production units.

(h) The focus must shift from conventional hardware to informalization in order to orient all defence technology to Digital and Network based.

Chapter 3

Defence Expenditure and Offsets

Military Expenditure

1. World military expenditure has gone up due to technological development as well as the way the future wars may be foughtThe expenditure is maximum in USAwhich accounts for almost 45 % of the world milit ary expenditure according to SIPRI. The other major countries are UK, China, France, Japan .India accounts for only 2 % of the World military expenditure. The Indian Defence budget has increased from Rs.1008 crores in 2007 to 1037 in 2008 and Rs.1066 crores in 2009.The details of the international trends are tabulated below.

International trend in military expenditure

in 2007 - in Billion Dollars

Rank	Country	spending	World share %
1	USA	547	45
2	UK	59.7	5
3	China	58.3	5
4	France	53.6	4
5	Japan	43.6	4
6	Germany	36.9	3
7	Russia	35.4	3
8	Saudi Arabia	33.8	3
9	Italy	33.1	3
10	India	24.2	2
11	South Korea	22.6	2

Source : SIPRI

Nuclear Power

2. In the field of military power , the deployment of nuclear war heads gives deterring power to a country. The leading countries in the field of Nuclear powers are USA, Russia, U.K, France, China, India, Pakist an and Israel. The number of strategic warheads, nonstrategic warheads and warheads deployed as estimated by SIPRI are as follows.

World nuclear forces-2008

Country	Strategic warheads	Non strategic warheads	Warheads deployed
USA	2575	500	4075
Russia	3113	2076	5289
UK	185		185
France	348		348
China	161	15	176
India			60 to 70
Pakistan			60
Israel			80

Source : SIPRI 2008

International Trade in Military Hardware

3. Imports: - India ranks as number 2 in import of military hardware after China during 2003-2007. China imports were 13,463 Million Dollars as compared to Indian imports valued 9105. Million Dollars

Exports: - The leading exporters of Military hardware during 2003 to 2007 in Million Dollars were USA, (34,499); Russia (28382) and China (2057).In contrast to the above figure Indian exports were only 43 .

4. China spends 2.1 % of their GDP on Defence while India spends about 2.7 %. In absolute terms the Chinese expenditure is more than double of India .In terms of import of defence equipments, China ranks number 1 followed by India in second place. Incidentally USA also import s some defence equipment but they are ranked at 10 th position. The international trade in defence equipment is growing and US tops the list with largest share. Russia is number 2 in export of defence items and China ranks at 8 th position.The details of the defence trade are given below which is based on SIPRI-2008 data.

Parameter	USA	CHINA	INDIA	OTHERS
Imports-20037 in Mn.$	2601	13,463	9106	Israel - 4239 Pak - 2347
Exports 2003-7 in Mn $	34,499	2057	43	Russia - 28,382

5. USA has been the largest exporters with 63.5 % share of the world market. The other leading countries are UK, Franc and Italy. The share of India was only 1.5 %.

Country	Number of companies	Arm Sales-2005 Billion $	Arm sales-2006 Billion $	% Share of top 100 arms sale in 2006
USA	41	183.6	200.2	63.5
UK	11	35.2	37.3	11.8
France	6	19.9	19.5	6.2
Italy	5	10.6	11	3.5
Germany	5	5.2	6.1	1.9
Russia	8	4.6	6.1	1.9
Japan	4	5.4	5.2	1.7
South Korea	3	1.7	1.8	0.6
Israel	4	3.7	4.6	2.9
India	3 *	3	3.5	1.5
* This includes Indian Ordnance factories, HAL & Bharat Electronics				

Source-SIPRI-2008

Offsets Concept and Utilization

6. International trade in armaments does not take place under normal market conditions. The Buyer and the seller sometimes have extended interest linked to this trade other than contracting for a product. Therefore a form of counter trade has emerged called offsets. There is an ongoing debate on the merits / demerit of offset worldwide. India has just begun the process and has to gain experience. Some countries have set up expert organizations to oversee the implementation of offset and advise the government.The fact of the matter is that come what may, in the world of business, there is no free lunch. Business will invest in any scheme as long as it tends to get a monetary gain else it's a no show. It would also be prudent to examine the experience of other counties on implementing offsets. Notable amongst them are Poland, Indonesia, South Korea, South Africa, Israel and Brazil who can be compared to India. The Defence Offset Facilitating Agency (DOFA) has been formed; its role, duties and staffing parameters are given below:-

DOFA has been established under the Department of Defence Production (DDP) as a single window agency to facilitate implementation of the offset policy of the Ministry of Defence. DOFA will assist potential vendors in interfacing with the Indian defence industry for identifying potential offset products/projects as well as provide requisite data and information for this purpose.

Charter of Duties

- Facilitate implementation of the offset policy.
- Assist in vetting of offset proposals technically.
- Assist in monitoring the offset provisions.
- Suggest improvements in the policy and procedures.
- Interact with Headquarters Integrated Defence Staff and Service Headquarters.
- Advise, in consultation with the Headquarters Integrated Defence Staff, Services and Defence Research and Development Organisation, areas in which offsets will be preferred.
- Promote exports of defence products and services.
- Provide advisory clarifications on the policy and procedures (in consultation with the Acquisition Wing wherever necessary).

Organisation

1. DOFA has been established under the Department of Defence Production (DDP) and functions under the chairmanship of Joint Secret ary (Exports). It is being serviced by the Directorate of Planning & Coordination, in the DDP under Director (Planning & Coordination) who is assisted by Planning Officer (Capital Acquisition Plan), PO (CAP) and Deputy Planning Officer (Capital Acquisition Plan), DPO(CAP).

2. Director (P&C) is the designated officer for interface with Industry and Acquisition Wing.

3. DOFA has been constituted as under :-

• JS (X), Department of Defence Production	Chairman
• Representatives of three Service HQrs (Indian Army, Indian Air Force, Indian Navy)	Member
• Representative of Headquarters Integrated Defence Staff (HQIDS)	Member
• Representative of Defence Research & Development Organisation (DRDO)	Member
• Designated officers from Hindustan Aeronautics Ltd (HAL), Bharat Electronics Ltd (BEL) Bharat Earth Movers Ltd(BEML) and Ordnance Factory Board (OFB) (on a case to case basis)	Member
• Designated person from Confederation of Indian Industries (CII) and Federation of Indian Chambers of Commerce & Industry (FICCI)	Member
• Director (P &C)	Member Secretary

Email address - dofa-mod@nic.in

The DOFA will assist potential vendors in interfacing with the Indian defence industry for identifying potential offset products/ projects as well as provide requisite dat a and information for this purpose. The agency will function under the supervision of a designated Joint Secret ary of the Dept of Defence Production, and have representatives from the Service HQrs, HQ IDS, DRDO, DPSUs and OFB as well as from CII, FICCI and ASSOCHAM and other agencies as deemed necessary. DOFA may set up committees and sub groups as considered necessary or based on the inputs received from DRDO.

DOFA may also engage, following a fair, rational and transparent process, reputed independent professional expert bodies, as suggested by DRDO, to assist in its functions and commission studies by such bodies on offset policies, their implementation, utility and impact.

As per available information DOFA is yet to start functioning in real earnest, it may be that there are no offers in the pipeline, yet, and it will start functioning as it gains experience. It would also not be prudent for DOFA to formulate polices without assessing the pros and cons of offset in the Indian context. Prima -facie no country has built up its industrial base solely using the offset route.

7. Before the effects of offset are debated, the fundamental question that need to be answered is "*what is the aim of offset that has to be achieved?*" Is it generation of employment, revive a flagging industry, for political prestige, or to keep national assets in the country. What is India's aim to use offset needs to be clarified. India is the second largest importer of military products after china and after 60 years of independence it has only 30 percent of indigenous content in the defence sector. Due to the economic growth India needs to safeguard its interests and needs a self reliant defence structure. In case this is not independent it will be detrimental to an independent foreign and economic policy India cannot afford to fritter away its achievement by having a weak defence industrial base. In fact this would be India's Achilles heel. Vested interests can tie down India by encouraging proxy war or insurgency in parts of the country. The Indian economy has been improving at a healthy rate of around 9 percent at an average during the past few years. The industrial output especially consumer goods are being consumed within the country. Remittances from abroad from knowledge workers and the Indian Information technology industry is steadily improving. These factors have contributed to increased employment in the country. As a result of improved economy in the region India has enhanced its image. Force projection by military operations outside the national borders (out of area contingency) has further improved India's image as a regional power.

It would therefore be reasonable to assume that India would like to develop critical technologies indigenously through the offset route to reduce foreign dependence for military hardware.

8. Independence of national defence can only be achieved by developing indigenous industry and not by relying on imports or offsets as brought out by Jurgen & Dunne in their work on "Arms trade and economic development: theory policy and cases in arms trade offsets" have convincingly argued on the reality of offset and they argue that:-

"An increasingly important facet of the international arms trade is so-called offsets, Arrangements that obligate the arms seller to reinvest ("offset") arms sales proceeds in the purchasing country. In justifying arms expenditure and in promoting local industrial activity offsets are claimed to offer significant benefits to developing countries, yet until recently there has been little research on how well offsets work in practice. Empirical evidence suggest that there is virtually no case where offset arrangements have yielded unambiguous net benefits for a country's economic development. As a general rule arms trade offset deals are more costly than 'off-the-shelf' arms purchases, create little by way of new or sustainable employment, do not appear to contribute in any substantive way to general economic development, and with very few exceptions do not result in significant technology transfers, not even within the military sector."

9. The concept of Offsets in defence purchase has been internationally used by many nations. USA, the largest exporter of arms in the world has offered Offsets in variety of ways. During the period 1993-2007, USA entered into 625 Offset agreements with 44 countries. The total value of the Ofset agreements was $65,460.58 million.The percentage of Offsets varied from 34.33 % to 124.92 % giving an average value of 71.89 %. The Offsets have been classified into three categories:-

- Direct offset which include subcontracts and co-production.
- Technology transfer, training, licenced production, oversees investment, credit assistance, miscellaneous.
- Indirect offsets which includes purchases.

10. The actual offset transactions for the diferent categories during 1993 to 2007 in terms of % were as follows :-

Transaction category of Offset	% age of total offsets during 1993-2007
Co-production	7.03
Credit assistance	4.22
Licenced production	0.77
Miscellaneous	7.37
Oversees investment	2.54
Purchase	37.00
Subcontract	22.3
Technology transfer	16.69
Training	2.08

11. It may be observed that maximum offsets were in the form of purchase, followed by subcontracts and technology transfer . (Source-Thirteenth Study Report to Congress on the impact of offsets in defence trade prepared by the US Department of Commerce, BIS, December 2008)- USA also set up an Interagency Working Group in December 2003 for continued dialogue on limiting the adverse efects of offsets in Defence procurement. The group consists of France, Germany , Italy, Spain, Sweden and UK who are the leading European arms manufacturing countries. The areas of concern are given in the Annual progress report 2008.

12. Offsets have been used by large number of developed as well as developing countries in purchase of defence equipments.There is wide variation in the policy followed by different countries for the use of offsets. The key parameters in offset policy are as follows:-

- Set minimum value of defence contract that would attracts offset provisions.
- Offsets required as % of the contract value.
- Sectors for which offsets would be applicable.
- Use of multipliers. Multipliers are used to give different weightage to diferent types of offsets. This gives direction in which importing country is interested.
- Banking of offsets against future deals.

13. Some of the countries which have a well defined offset policy are Australia, Canada, Finland, Greece, Israel, It aly, the Netherlands, Norway , Poland, South Korea, S pain,

Switzerland, Taiwan, Turkey and UK. Summary of the policies followed by them is given below. (Details are given in Appendix- A)

Ser No	Country	Minimum Value of Defence Contract	Minimum Offset Required	Offset Sector	Multiplier
1	Australia	US $ 3.75 Mn	No specific Min/ Max	Defence	None
2	Canada	NA	100%	Defence & Civilian	None
3	Finland	NA	100%	Defence	0.3 for exports Of finish Products. For others multiples are negotiated
4	Greece	10 E million	120%	Defence	Upto 10
5	Israel	US $ 0.5 million	35%	Defence & Civilian	1-1.5
6	Italy	US$ 6.6 million	Not < 70%	Defence	Max of 3
7	Netherland	5 E million	100%	Defence & Civilian	Negotiable 1-5. 5-10 and 10-30
8	Norway	US $ 6.7 million	100%	Defence & Civilian	0-5
9	Poland	5 E million	100% (Defence 50%)	Defence & Civilian	Negotiable up to 2-5%
10	South Korea	US $ 10 million	30%	Defence	Determined by Authorities
11	Spain	NA	100% but may vary	Defence & Civilian	Between 2-5 when used
12	Switzerland	US $ 17 million (may vary)	100%	Defence & Civilian	Max of 2-3
13	Taiwan	US $ 10 million	Will increase to 70%	Defence	1-10
14	Turkey	US $ 10 million	50%	Defence & Civilian	1-5
15	UK	US $ 17.2 million 50 Mn Pounds for French & German Companies	100% target	Defence	No multiplier for IP credit

Offset Issues in India

14. India at present is the second biggest importer of military hardware in the world. In the coming plan period India expects to spend more than a US $ 100 Bn for arms procurement. India has recently initiated the process of adopting the offset route to establish its defence industrial base (DIB). It is likely to attract about Rs 45000 crores in of fset investments. Further this is likely to extend into the 12th plan also. Therefore a vision of the next 10-15 years has to be taken when forming Policy or t aking investment initiatives. The scope of business in Indian defence industry is immense.

15. While the Indian government has initiated the process of employing offsets, the views of the Vendors and suppliers in the private sector on the subject also merit consideration. The views expressed by them in the Conference organized by IDSA in Oct.2008 .These are summarized below as projected by Boeing, BAE, M&M, L&T, Wipro, Tata-Power and Lockheed Martin.

Views of Private Sector on Offsets in Indian Defence Acqusitions

- √ There is a need for transparency of potential order book of products which should be shared with the Private sector.
- √ In DPP 2008 the time for RFP is minimum and needs to be enhanced.
- √ The Qualitative requirements (QR) for Products must not be overstated the last 10% of the requirement is likely to cost majority of the cost therefore the QR has to be reasonable.
- √ In certain case the Government will have to fund R&D.
- √ India should identify its place in the global market of defence Industries, whether it wishes to be a market leader, follower or Niche player and also the technology that it wishes to focus on.
- √ Incubation of technology in emerging areas must be initiated to be ahead of the pack.
- √ Relationship between defence &Industry should be established as partners and Government should not behave like a buyer.
- √ While DPP 2008 has been formed the process remains the same. Example - the concept of L1, LCC, RFP and NC- NC (No cost - no commitment) are still there. These issues need to be addressed.

√ The DPP 2008 is not adequate for procurement of $30 Bn.

√ Entry to international supply chain is difficult but it can start with labour intensive products and move up the ladder into higher end products.

16. **The Emerging Challenges are:-**

- Identify where are the competitive advantages? Select niche markets; understand the 'sweet spot'.
- Aggressive development of Supply Chain Management (SCM) by PSU/DGOF . Indigenous supply of raw materials
- Obtain Aerospace certifications, for related industry.
- Getting embedded in global supply chain by entering into JV & acquisition overseas which is promising at present when their capitalization value is low.
- In offsets there must be a 'win-win-win' situation for Government, OEM and Local Industry.
- In offset the focus must be on two things, first the Policy objective and second, Evolving offset guidelines. This forms a frame work and must be revised regularly
- Commercial and defence requirement s may be clubbed together so that of fsets can be provided together, in the Boeing experience the aero technology in critical areas of commercial and defence are same.
- Offset structure – There must be an organization dealing with offset for industry to contact. This point must be able to interpret the offset policy in a consistent manner and guide the investors.
- The acquisition team and the Offset teams should be different so that they do not distract each other. Thus the OEM can focus on Offset guidelines. The offset levels will not vary and will be even for all. It will be a single point of cont act and as priorities of requirements emerge they will work with the OEM to add new priorities.
- Banking guidelines for of fsets must be reviewed regularly and a single authority must handle it.
- LM has entered into projects with Korea – aerospace precision castings, Belgium – composite materials, Taiwan – inertial navigation systems, repair and overhaul,

Indonesia –Training, F16 software, maintenance and training with several countries, Turkey – TUSA aerospace industry and parts.

- Attributes of a successful offset program.
- Are the goals of the country clear? Avoid wheel spinning.
- Provide flexibility to offset partner. Rigid rules increase challenges and identity to change to changing circumstances.
- How broad is the scope of Offset.
- Should it include internal security / are conditions consistent with the objectives.
- What is the level of investments permitted (in India it is only 26%).
- The offset must make business sense to both parties.
- Recognition that Offset adds to the cost. There are no freebies.

17. The Government policy for keeping FDI level at 26% ostensibly is to ensure that the foreign collaborator has a say in the major Board decisions since it requires 75% representation for approval, further Of fset is a means to an end and not the end it self hence the focus of the collaborator has to be on the main business at hand. Technically there is little dif ference between 26% and 49% except the margins of investment and profits. The fears are probably based on assumptions that this may facilit ate a hostile takeover by the foreign collaborator in the future by acquiring just 3% more to hold the majority share. This is due to a bias in the Indian context, because of p ast experience of the technology denial regime, and will take time to get over as mutual confidence is built up. Therefore as a starter, it must not be a matter of undue concern.

18. While offset terms may be negotiated it needs a corresponding cap ability to absorb the plans by way of a knowledge base and functional organizations. This requires availability of manpower, technical capability and Infrastructure, in that order, to absorb and upgrade the new technology. There are other imperatives like national laws, repatriation of profits, coordination of effort between intra government agencies etc though not discussed here. Manpower has been discussed separately in this study and the infrastructure can be built up. How to absorb the technology and to set their priorities is of the essence. Sujeet Samaddar (Defence development and national security) has convincingly argued the case for absorbing technology, identifying the sectors where state of the art technology must be sought and the Short, Medium and Long term plans formulated.

19. It is therefore apparent that India cannot unilaterally develop its DIB to a required level of self reliance. Technology is not easily parted by those having it. Therefore self help is the long term answer. In the short term offsets may be employed to leverage our strength being a purchaser to *obtain some critical technology* not available in the country and thereafter develop it further (therefore the selection has to be deliberate).

20. Dr.Prahalad & S.Radhakrishnan from DRDO have strongly recommended the leveraging the large purchasing power through offsets to procure denied technologies and specialized facilities into the country in a relatively shorter time. Represent ative set of critical technologies and specialized facilities suggested are as follows:-

a. Missiles related technologies are:-

- Uncooled FPA seeker for PGMs.
- Multi Disciplinary design optimization.
- Optimal/ adaptive control systems.
- MEM base Dual Mode seekers.
- MMW based imaging seekers.
- High temperature aero-structures.

b. Aeronautic related technologies are :-

- Airframe shape optimization.
- Re configurable control systems.
- Multispectral data fusion.
- Multiband flexible conformal antennas.
- Adaptive / optimal / model reference control
- Thermal barrier coatings.
- Manufacturing of single crystal turbine blades.
- Surge margin improvement (casing treatment) in gas turbine engines.
- High accuracy direction finding using phase interferometer (I degree).
- Aerodynamic design and shape optimization for Aerostats.

- Autonomous landing, take-off and navigation for UAVs.
- Smart aero-structures.
- Airframe engine integration and optimization for supersonic air craft.
- Aero data prediction package "Digital DATACOM".

c. Materials related technologies are:-

- SIC / Si Cr technology.
- Ultra high temperature ceramics such as ZrB/ ZrC / SiC for leading edges.
- Development of high temperature steels, Ti alloys, Al alloys and advanced ceramics such as B_4 C and Ti B_2.
- Gun barrels.
- Light weight ballistic materials phase change materials carbon and inorganic nanotubes.
- Carbon monofibres and nano-coils.
- Metallic W, Nb, Ta , Ti alloys and structures.
- Non metallic composites / MMC/ Multifunction materials.
- Polymers.
- Chemicals –radar absorbing materials.
- High energy materials (Nano materials and Endothermic fuels).
- Propellants and fuels.

d. Technologies related to the Naval systems are:-

- Super Caveat technology.
- Pump jet propulsion.
- Non acoustic detection.
- Active vibration control.
- Air independent propulsion systems.

e. Technologies related to Life sciences

- NBC Defence technologies.
- Underwater escape system for depth greater than 100 meters.
- CNT base sensors.
- MIP base sensors.

f. Specialized facilities related technologies:-

- High enthalpy facility.
- Hypersonic wind tunnel.
- Free piston shock tunnel.

Offsets: International Experience

Turkey

21. Turkey signed 66 offset agreements during the period (1985 to 2007).Policy for Offsets was formulated in 2002. It is applicable to contracts valued more than 5 million Euros from one supplier spread over 3 years .They have implemented 20 agreements so far. The Offset obligation is 50 % and penalty of non fulfillment is 6 % as per their current policy. They use different multipliers for giving credits. The total value of OFFSETS was -6.1 Billion dollars. Value of Fulfilled offsets is 3.3 Billion dollars which is approximately 55 % of the total offsets. They signed Offsets agreements with Germany, USA, UK, Italy, Spain, Israel, France & the Netherlands. Major benefits gained by them are:-

- Software development for helicopters
- Export of helicopter parts
- Investment in export oriented defence industry
- Export of components
- Became preferred supplier of components to Lockheed

Poland

22. Poland formulated the Offset policy in 2002. The signed 12 offset agreements during 2001 to 2007 for total value of 6.78 billion Dollars and 1.1 Billion Euros. The Areas for offsets were:-

- Transport aircrafts
- Light torpedoes
- ORKAN class ships
- F-16 aircrafts
- Patria vehicles
- Gun turrets
- Ammunition
- Unguided rocket systems
- Elements of weapons
- Antitank missiles
- Digital radio stations
- Missiles
- Gas turbine for ships

Learnings from Experience of Poland

- Need to define role of offsets clearly
- Negotiations are very important aspects of Offsets
- Differentiate offset multipliers
- Preferred areas for offset are technology, investment, R&D,

Korea

23. Korean started using Offsets fom1988.Korea has been passive in promoting exports through Offsets. They give priority to technology acquisition through Of fsets, especially those technologies which are not available otherwise. According to them the Valuation of Offsets is not easy. Their experience in this area is summarized below.

- Indigenous manufacture through local vendors promotes further subcontracting to local companies.
- In mature industries / products, where the product has been traded internationally many times, the Offset opportunities get depleted.
- Korean system does not recognize foreign investment as an Offset.
- Offsets have been very useful in international R&D cooperation e.g.
  - Concept development of T-50 Trainer as KPF Offset.
  - Technology transfer on KFX fighter as a F-15K Offset.
- Actual cost borne by foreign supplier to fulfill Ofset trade is ~2.7 % of the contract price.
- Usefulness of Offset depends heavily on the negotiating skill.
- Korean Offset policy is applicable to contract s more than 10 million dollars.
- Korean exports due to Offset is ~ 14.9 % of their exports during 1983 to 2005.

24. Korea has used Of fsets for Technology transfer –know how, Technology transfer – others, Export s, Equipment. The major use of of fsets was in technology transfer and equipments. The utilization of offsets for different purposes is given below

Item	Value of Offsets In 100million dollars	%
Technology transfer-know how	27.0	44
Technology transfer -others	5.5	9.2
Exports	13.1	21.8
Equipment	14.5	24.1
Total	60.1	100

25. Korean experience and lessons learnt are summarized below:-

- Offsets have helped to develop defence industry
- Off set valuation is difficult
- It is likely that extra payment is charged for the Offsets
- Offsets promote international cooperation
- Offset valuation should be done by the supplier and the buyer should have the option to cut-off price as a condition for exemption from Offset obligation
- Prefer technology related to R&D in progress
- Use Offsets for explorative R&D also

Recommendations

26. The recommendations are as follows:-

(a) Considering India's defence expenditure plans it has an inherent leverage which must not be lost and should be the focus in each purchase to obtain technology, Process or Services.

(b) There is a very strong case of establishing a functional facilitating agency for which resources may be outsourced if required as a single point of advice. DOFA is in the nascent stage and needs to be an independent agency. It must include capability to carry out market research of defence industry. It should also compile data / studies on Indian offset experiences.

(c) The policy on Offsets should be reviewed based on experience of its application at regular intervals.

(d) Transparency is of great value in making utilization of offsets a success and entails giving complete project information, selection criteria, future expansion, time frame and other liabilities in the RFP stage.

(e) The time frame for according an approval or Rejection must be fixed and is not open ended.

(f) Large companies contracting Offsets should be encouraged to farm out small portions to the SME in order to strengthen the DIB.

(g) The investment level of 26% must be maint ained till experience and mutual confidence is established. This must be viewed in the light of emerging realities of global recession, increasing mutual interdependent world, India's promise of further business and the fact that foreign investors need India as India needs them (the world is growing flat). The foreign investor need have no fears if they intend to honour their commitment in the long term.

(h) The investment level could be raised up to 49% in case where the supplier intends to establish manufacturing facilities from scratch. A conditional clause could be added in the policy that on successful completion of their of fset commitments an investor may apply for enhancement of their share with evidence of future business potential. This would be assessed on case to case basis. The government must make it mandatory to supply the reasons for approval or rejection.

(i) The subject, again, underlines the necessity of forming a formal long term security policy and consequent acquisition plans.

(j) Offset trading/ transfer could be made transferable subject to the transferee using them for defence industry only.

(k) The list of areas where of fsets are required should be evolved jointly by users, CIDS, DRDO and other Research agencies. It should be revised regularly to make it more specific and to ensure that an investor 's proposal is not shot down at the eleventh hour.

(l) The need to apply multipliers in giving credit may be examined to give direction for flow of offsets in the desired direction. The terms of reference should be to draft a framework of a win-win situation.

(m) The Research agencies in the country may be brought into the offset loop to develop dual use technology.

Chapter 4

Present Status of Defence Industrial Base-2009

1. Defence Industrial Base (DIB) includes the capability for production, R&D, manpower development .The growth of DIB depends heavily on the policy environment .In this context, the policies which affect the capabilities for building the DIB also need to be examined. The present status of the various components of DIB is examined below.

Defence Production

2. A large defence industrial base already exists in India. It consist s of the following components:-

- Ordnance factories.
- Defence public sector enterprises.
- Other public sector enterprises.
- Private sector including small scale and ancillary units which are suppliers to Ordnance factories, PSE's, and other defence establishments.

3. The overall defence production from the different components is given below:-

Defence Production in India in Rs. Crores

Year	Defence production-Ord. factories	Defence Public sector*	Total defence production
2007-8	3050.80 upto Nov.2007	6382.3	9433.1
2006-7	6197.35	15489.3	22046.65
2005-6	6891.69	13025.07	19916.75

* This includes 8 Units, namely HAL,BEL,BEML,MDL,GRSE,GSL,BDL & MIDHANI
(Source -Annual report of MOD 2007-8)

4. The Status of the Ordnance factories is as follows: -

Ordnance Factories: - This is the largest departmentally run organizations in the country with 39 factories at 24 locations. Their turnover in 2006-7 was Rs. 6197 crores. Indian Ordnance Factories have been ranked as 45th in the list of top 100 defence armaments producing companies (turnover of 1589 Billion Dollars)by SIPRI in2006.Sixteen factories were set-up before independence and the balance between sixties and nineties. Factories which came up afer 1970's gave up the previous practice of in-house backward integration, and stress was given to production of finished products with supplies coming from outside sources. The product range of the ordnance factories include the following range of products for the armed forces:-

a. Arms-small arms, tank guns, rocket launchers.

b. Ammunition systems for weapons, rockets, missile warheads, grenades, bombs for Air force and Navy.

c. Engineering equipments –armoured and transport vehicles, tanks t-72, T-90, infantry combat vehicles etc.

d. Opto-electronics instruments / devices / fire control, for armoured. vehicles , infantry and artillery systems.

e. Troop comfort items like clothing etc.

5. They have a large pool of qualified and experienced personnel with overall strength of around 1, 12,000. They have a wide network of training institutes for all categories of employees. Total Quality Management (TQM) concept has been given major thrust in all the ordnance factories. They have small set up for product and process improvement. They have developed some ammunition items and electronic system for the 155mm guns.

6. The Ordnance factories operate primarily on the basis of orders received from their customers and they hardly take any project on their own initiative, though they have the potential for this. They are capable of competing with exports to some extent if certain systems are introduced for forecast demand for their product. They can develop or produce through technology transfer mechanism. In the current status they do not appear to be taking such pro-active role. Their users often complain about their deficiencies regarding competitiveness in terms of :-

- Cost
- Delivery

- Quality

7. The above factors result in import also though the cap ability exists within the organizations to meet the requirements of the customers. The systems related to internal and external operations may make them more competitive in future. At present 70 % of the armaments are imported. The calculations base on SIPRI data for 2oo3-2oo7 shows that the imports are 85 %.

The status of the productions needs to be examined from the technology life cycle concept. Attempt has been made to estimate the technology life cycle status for the major product groups as perceived by the users. Graphical summary of the same is given below-

Technology Life Cycle Status of Defence Production in India, DIB-2009

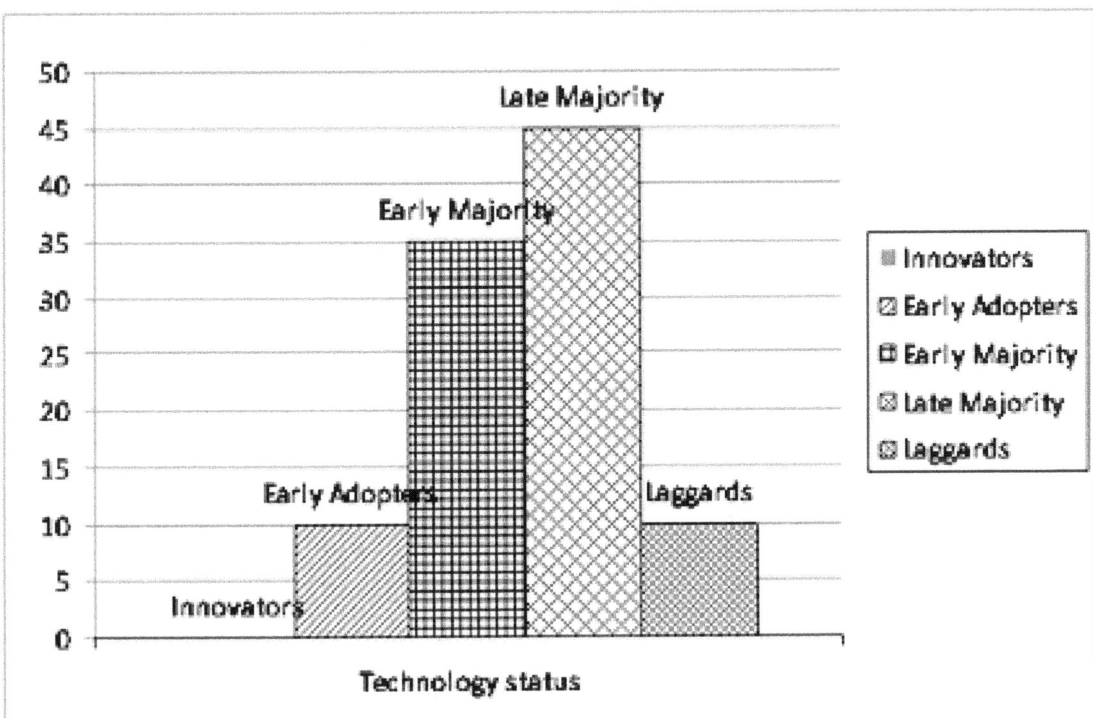

The conclusion of the analysis is that the innovative stage is conspicuously absent. Few of the products are in the early adopters st age and bulk of the production is in early / late majority .10% of production is in Laggard stage using outdated technologies.

8. The ordnance factories operate primarily on the basis of orders received from their customers and get mostly involved with import substitution .For meaningful contribution, they should be proactive to forecast the requirements of their customers and develop items in time to meet their requirements and hence increase overall dependence on indigenous production. They can obtain technologies through technology transfer from abroad, develop it themselves or take technologies from an indigenous R&D institution. There is also scope for rationalization of their operation through BPR .So far Ordnance factories have operated in monopoly environment and now since the private sector has been permitted / encouraged to enter into defence production, Ordnance factories should concentrate on items where they would have the competitive advantages. This may call for major reorganization in the operation of the ordnance factories.

9. Ordnance factories have experience in production of arms, which can be fruitfully utilized Research and Development in their field.They should have a proper R&D set up of their own up gradation of existing products , absorption of imported technology and product development .They should become the main source of futuristic armaments R&D could be a part of the existing factories or through holding company structure, in which their R&D could be in a separate unit under the control of the CEO, who controls other factories

10. Public Sector: - There are 8 Public sector companies under the Defence Ministry which manufacture large variety of defence equipment. Their production in Rs. Crores in the last three years was as follows:-

Name of PSU	Production In 2004-5	Production In 2005-6	Production In 2006-7
HAL	4984.55	5916.62	9201.88
BEL	3234.97	3449.74	4012.76
BEML	1885.95	2179.74	2590.75
MDL	540.63	518.37	1872.24
GRSE	470.28	662.18	641.66
GSL	141.83	249.78	267.07
BDL	465.79	534.28	385.84
MIDHANI	141.67	177.60	223.88

Source-Annual Report of MOD, 2007-8

11. Brief on their capability is as follows:-

Hindustan Aeronautics Ltd (HAL). They have 20 production unit s, 9 R &D centres and 4 joint venture s at present. They employ 31,666 persons. Their R&D expenditure is approx. 6.7 % of their turnover . They have so far produced 1 1 types of aircrafts from in-House R &D, and 14 types of aircrafs under licence with foreign technology Indigenously designed aircrafts in current production are :-

- Advanced light helicopter (ALH)
- Intermediate Jet Trainer (IJT)
- Light Combat Aircraft (LCA)

The major products under licence are:-

- SU-30MK1 fighter aircraft
- Jaguar
- Hawk
- Dornier (DO-228)
- Cheetah and Chetak helicopters

HAL will be participating in design and development of Multirole transport aircraft and fifth generation Fighter aircraft. They are also partners in Space program and have diversified into Industrial and marine gas turbines and real-time sof tware areas. In the year 2006-7 they exported goods worth Rs. 270 Crores.Their indigenous content is stated to be around 74 %.

Bharat Electronics Ltd

12. They are engaged in design, development & manufacture of electronics equipments and components for the use of Defence services and p aramilitary organizations. They have 7 production units spread over 7 states. More than 57 % of their products are based on their in-house development.They spend 5 to6 % of their turnover in R&DTheir product range includes:-

- Radar for defence and civilian use
- Sonar products
- Communication equipments, HF /VHF/UHF

- Microwave bands for tactical and static application.
- Telecommunication equipments.
- Broadcasting equipment.
- Night vision devices.
- Electronics components including IC's, ASIC's, hybrid microcircuits.

13. They are currently working in the following technology areas :-

- Frequency hopping radios.
- Encryption.
- Software defined radios.
- Mobile satellite terminals.
- C4I Systems.
- Phased array Radars.
- Airport surveillance radars.
- New generation Sonars.
- Electro-optical fire control systems.

14. BELis also diversifying into the areas of satellite base system solution, solar photovoltaic systems, smart card base systems, X-Ray baggage and cargo inspection system, compact vacuum interrupters, set top boxes etc.

BEML Limited

15. BEML is engaged in design and manufacture of wide range of construction, mining, defence, railway and metro coaches. The Government equity in this company is 54.23 % at present. Three of their unit s have ISO-9001-2000 certification. They have global operations in Malaysia and Brazil.They export their products also. Exports in 2006-7 were Rs. 110.73 Crores.They have R &D sett up also in which they spend 0.7 % of their turnover

Mazagaon Dock Limited (MDL)

16. They design and manufacture naval equipment. They are ISO-9001-2000 certified company, and the product range includes:-

- Destroyers
- Frigates
- Missile boats
- Corvettes
- Submarines (Scorpene class of French design)
- Patrol vessels
- Merchant vessels and dredgers, fishing boats, multipurpose ships
- Offshore platform & sub-sea pipelines

Goa Shipyard Ltd.

17. They build medium sized vessels for Indian Navy and Indian Coast Guards and others. They have ISO-9001 Certification and their product range includes:-

- Offshore patrol vessels of large variety
- Missile boats
- Survey vessels
- Sail training ships
- Torpedo recovery vessels
- Passenger vessels and tugs etc.
- Stern gear equipment with collaboration of France

They have a plan for modernization and expansion of their facilities by three times.

Garden Reach Shipbuilders &Engineers Ltd. (GRSE)

18. They are engaged in manufacture of ships and complex engineering items. Their product range includes

- Frigates
- Corvettes
- Hovercraft
- Water jet fast attack crafts

- Engines for naval and stationary applications.

Bharat Dynamics Ltd. (BDL)

19. They have the capability to produce state of the art Missiles. They have three production units with CAD / CAM facilities and their product range includes

- P-II Missiles
- Konkurs-M and Invar with Russian collaboration
- Counter measure dispensing system for Air force
- Systems / subsystems for missiles like AKASH, NAG, AGNI
- Light advanced torpedoes
- Decoy launching system
- Light weight UAV s
- High speed heavy weight torpedo (under development)
- Propellants for missiles

Mishra Dhatu Nigam Ltd.

20. Their fields of operations are specialized metal / alloys products required in strategic sectors like, Aeronautics, Space, Armaments, Atomic energy, Navy, DRDO. Products include Molybdenum wires and plates, Titanium and stainless steel tubes, seamless tubes, alloys for electrical/ electronic and magnetic applications, controlled expansion alloys and resistance alloys etc. They are in the process of modernizing their plant .They have an active R&D programme and spent 1.14 % of their turnover in 2006-07.

Private Sector in Defence Production

21. The private industry was not allowed to manufacture Defence equipment as the same was reserved for manufacture by the Government till 2001. Some of the units supplied some of the components and fabricated items for defence units .In 2001 the Industrial Licensing policy was liberalized to permit private sector with FDI up to 26 %. Till 2007 73 letters of intents have been issued to various Indian Companies. The list includes some of the well established organizations like BHEL, Tatas, Mahindra & Mahindra, Larsen & Toubro, Samtel. Some of the major products included in the approvals are:-

1.Underwater weapon systems	2. All types of guns
3.Armoured and combat vehicles	4.Communication systems
5.Propulsion systems for ships	6.War ships, submarines, high speed boats
7.Radar and sonar systems	8.Weapon launchers
9.Torpedo land / naval mines	10.airborne assemblies for aircrafts, UAV and helicopters
11.mobile surveillance systems	12.night vision devices
13.military aircrafts and avionics	14.network centric warfare enablers
15.GPS based navigation and tracking systems	16.Guided munitions
17.Smart ammunitions	18.Reflex sights and aiming devices
19.Radio and satellite communication equipment	20.Electronic warfare systems
21.Guns, rocket and missile artillery systems	22.Multifunctional displays
23.Gyros	24.Aircraft engines

22. The Industrial licenses are at various stages of implementation. The major steps are the conversion of the letter of Intent into Industrial Licence af ter approval for the foreign collaboration, acquisition of land, capital goods and commencement of production. There are large number of small and medium sector comp anies who supply component s, raw materials , spares and services to the defence industry . One data base has been prepared by DGQA and second data base has been prepared by DRDO in Dec.2008.The Data base give the location and the areas in which they have / can supply product s to the defence sector.

23. The changing threat from an asymmetrical war and terrorism has drastically changed the role of the army . The efficacy of the armed forces depends upon the weaponry and other material required to perform its assigned role. It is therefore dependent on the capability of the industrial base of the country. It is important that industries that are aligned with the technology required in the future by the defence forces are nurtured and encouraged in every way in order to enhance the indigenous cap ability. This includes R&D institutions,

academic research, private sector as well as the government owned public sector enterprises (PSU). It also devolves on virtually all government ministries to examine which part of their function strengthens the defense forces of India. The face of warfare has changed and the defense forces are required to meet the external challenge and also contain terrorism and other anti national elements (ANE) in asymmetric conflict. Simultaneously it is also required to protect its assets beyond the borders and in space. Thus there is a need to build the strategic capability on land, Sea, Air and Space, as an emerging economic and regional power to protect its trade, commerce and energy needs in the future. Further, the military commander now operates under the lens of the press and international gaze, where collateral damage has to be kept to a minimum. This requires a very-very high degree of C4I3 coupled with high mobility and precision weapons. Naturally the DIB has to be aligned to develop the required technology and production base in order that the defence forces can effectively meet these challenges.

24. The challenges to security has brought into focus new concepts in modern warfare like RMA, network centric warfare, management of EMP, dominance of Space, Precision-munitions, stealth technology and anti missile defense. These coupled with core research in nano technology, biotech and cyborg are the future force multipliers that will help a country to project its war fighting capability and the ultimate aim of forcing an enemy to submit to its will without firing a shot. Procurement of material for the defence is a continuous and critical function. For the purpose of this study when we refer to defence procurement it shall mean items of weaponry and other special equipment not normally used by the non-military consumers. It assumes special significance since the source of supplies are limited, even if available within the country. The existing sources abroad also cannot be taken for granted in the hour of need. This raises a serious concern for the defence organization as operations are hindered for lack of warlike items as well as their spares. The example of a war being lost for want of a nail cannot be more appropriate, though exaggerated.

25. The genesis of the DIB in India starts from the pre-colonial era when a few Ordnance factories were established by the British in support of their forces operating in the middle-east and far-east. Post independence the department of defence science was formed to be later reincarnated as DRDO in its present shape. The charter of the DRDO was to carry out research and develop high technology items for use by the defence services. The actual production of successfully developed products was to be carried out by the DGOF. Both are separate entities and have their advantage and disadvantages. Further post independence our leader formed the Industrial policy of 1956, based on the Soviet model, wherein items of defence were reserved for production by the public sector only.

26. It has now been realized that even after sixty years of independence the defence industry has developed only 25 to 30 percent in the country. The emerging complexity of warfare highly dependent on technology has raised serious concerns on our ability to maintain a credible defence.This has prompted the government to take a string of measures to build the indigenous defence industrial base. These are:-

- Change the Industrial policy of 1956 permitting 100% private investment in 1991.
- Create a Defence acquisition agency for fast processing of the procurements.
- Establish the Defence procurement Procedure (DPP 2008) for capital acquisition and the Defence procurement manual (DPM) for Revenue procurements.
- Route all procurements through the Integrated Defence Staff as a single coordinating agency for approvals.
- Offset has been accepted as a route to develop industry.
- As an incentive Offsets can be banked for a period of two years.
- The Indigenization of products related to maintenance requirement has been delegated to the army (EME).
- Full financial powers have been delegated to the heads of DGOF factories.
- Inspection and quality control procedures have been simplified and streamlined.
- Above all government has agreed to revise the policy from time to time as experience is gained.
- Foreign investment has been permitted in the defence sector up to 26 percent.

27. The DRDO and the DGOF &DPSU'S along with the user have between themselves identified and developed a large number of sources for the supply of raw material, components, small assemblies and low technology items.As per the collective agreement of all agencies it is estimated that the total indigenous content has been developed up to 30 percent.The balance 70 percent consists of major weapon platforms and high technology spares. The DGQA has developed around 3000 sources pan India and the DRDO has reportedly nurtured and developed approximately 550 high technology sources, in addition to this are the vendor list of the User i.e. ,DGOF,DPSU's. Consolidated list of manufacturers and suppliers along with their capability has not been compiled so far. This makes a large potential base in India.There is a need to create an updated data base for this by MOD or an industry association.

28. Some of the reasons for lack of defence industry in the private sector are:-

- Lack of R&D facilities and investment in India.
- Lack of cutting edge technology.
- The defence products have a niche market in which demand is highly erratic.
- Exports were not allowed.
- While development was attempted and succeeded the development cycle was so long that the technology became obsolescent.
- Private sector was not allowed in this sector till 2003.
- Complicated procedures and red tape to induct or sell an item to the defence.

29. The paradox that has been highlighted post 1991 liberalization is that given an opportunity Indian talent can rise to the challenge and deliver. Post 1991 liberalization ,within a span of 15 years the private Industry has scaled international levels and companies are not only competing in the world but acquiring manufacturing and business assets abroad. A feat that was unthinkable 15 years ago. What has caused this transformation? It is no miracle but a simple unshackling of governmental controls. This holds great promise for the defence industry. There are different levels of production related units in India. These are :-

- Level 1 – the provider of raw materials and small components.
- Level II – Providers of sub assemblies and low technology products.
- Level III – Manufacturers of complete assemblies and equipment that is ready to use.
- Level IV – System Integrators.

30. While the Systems Integrators are the highest level, given the availability of the market and liberalized rules most others can integrate forward or backwards. This was one of the path taken post 1991. There are companies operating at level IV in India like the TATA Industries, Mahindras, Kirloskar industries,Ashoka Leyland, L&T, Reliance industries etc who are world class manufacturers and system integrators however their area of focus has been restricted to commercial non defence products. Some of these companies are now diversifying into the defence sector that looks promising from the commercial angle. The potential for business is large as assessed by N.Neihsialin his study 'Entry of private sector to defence industry: opportunities and obstacles' given below:-

Financial Projections of likely market size (All figures in billion of Indian rupees)							
Years	*2008-09*	*2009-10*	*2010-11*	*2011-12*	*2012-13*	*2013-14*	*Total*
Total budget with (+10%)*growth	1056.00	1161.60	1277.76	1405.54	1546.09	1700.70	8147.68
Capital budget (45.46% of total)	480.06	528.06	580.87	638.96	702.85	773.14	3703.94
Revenue Equipment portion =9% ?of total budget	95.04	104.54	115.00	126.50	139.15	153.06	733.29
Total(Capital 45%+9%) out of Revenue portion	575.10	632.61	695.87	765.45	842.00	926.20	4437.23
Share of domestic suppliers (46%)	264.54	291.00	320.10	352.11	387.32	426.05	2041.13
Foreign suppliers (54%)	310.55	341.61	375.77	413.35	454.68	500.15	2396.10

Notes :*

(i) This assumption is based the average annual increase of over 8% during the last six years. It is seen that the total aggregate market size available up to the end of five year in the defence sector for procurement of stores (excluding provisions etc) and equipment (capital plus revenue) is about 4437 billion rupees.

(ii) While this is the approximate estimated market size of the defence sector , the market size in the non-defence sector is equally significant. The annual sale of Ordnance factories alone to MHA is about Rs.500 crores and the Kelkar Committee estimated its to be Rs.3708 crores in the next five years i.e. up to 2010..(p.20April 2005 Report.

The author further goes on to state that the market share of private industry in India in the next five years is likely to rise from INR 149 billion in 2008 to INR 151 billion in 2014, this is incentive enough in a falling global market for the determined investor to esablish himself in the defence industry in India. The details are as follows:-

Market share of Private Sector (in billion rupees) 2008 - 2014							
Years	2008-09	2009-10	2010-11	2011-12	2012-13	2013-14	Total
From Domestic Supplier	79.36	87.30	96.03	105.63	116.20	127.82	612.34
From Foreign Vendor	69.87	76.86	84.55	93.00	102.30	112.53	539.12
Total Share of Private	149.23	164.16	180.58	198.64	218.50	240.35	1151.46

The above is however subject to the following conditions:-

(a) Private sector may not be able to make much headway in the areas of integrating systems within the next five years due to heavy investment & long gestation period.

(b) The government organization will tend to favour the public sectors/government agencies at various administrative levels for sometime to come particularly in comparable situations

(c) Private sector may corner large portion of defence market in r/o dual use items.

(d) Private sector will make fast headway where item has a dual use especially information technology related products. The same holds true for mature technology such as transport, communication and electronics.

(e) Offset policy shall facilitate the Private company to exploit quickly at least the commercial aspects of it though the foreign partners may not part with high-end technology in haste.

(f) Private industry may contribute to be an out source for the foreign supplier and may collaborate in R&D resulting in a reduction of the suppliers share to India. this could be shared by him with the government.

31. While the existing Integrators have already forayed into the defence Industry by producing Rocket launchers, Avionics, Torpedoes, Combat vehicles and high technology electronic items the world leaders in Information technology like Infosys and Wipro will not lag behind to enter into software design for defence systems and design embedded technology. Till 2008 it has been reported that more than 140 companies have been granted letter of Intent/ Industrial licence for the manufacture of defence products.

32. The comparative production in the Ordnance factories and the PSUs shows that the share ' of PSUs is approximately 70 % of the total production at present. The details of the same are given below:-

Defence Production and Imports-in Rs. Crores

Year	Ordnance factories	Public sector*	Total defence production	Exports	#Imports Million $
2007-8	3050.80 up to Nov 2007	6382.3	9433.1	342.46	24,249
2006-7	6197.35	15489.3	22046.65	NA	23,615
2005-6	6891.69	13025.07	19916.75	NA	22,273

Import data is for calendar year .Source : SIPRI -2008

* This includes 8 Units, namely HAL,BEL,BEML,MDL,GRSE,GSL,BDL&MIDHANI (Source –Annual report of MOD 2007-8)

Defence R&D and Technology Management in India

33. Highlights of the R&D Scenario in India are as follows:-

- National expenditure on R&D has been increasing steadily and has increased from Rs.18,088 crores in 2002-3 to Rs.37,778 crores in 2007-8
- In 2005-6 the R&D expenditure was approximately 0.89 % of GNP in India
- Investment in R&D by the private sector has doubled from Rs.3498 crores in 2002-3 to Rs.7444 crores in2005-6 and expected to reach Rs. 11,192 crores in 2007-8
- Major expenditure in R&D is by the Government ~74.1 % and 25.9 % by the private sector (2005-6). The compounded rate of growth of private sector R&D was 28.6 % as compared to 14% in the Central sector
- In the central Government 86 % of the expenditure was incurred by 12 major Scientific agencies of the Government , namely CSIR, DRDO, DAE, DBT , DST, DOS, MOES, ICAR, ICMR, ICT, MOEn and MNRE.

The share of DRDO was highest among the 12 agencies at 34.4 % in 2005-6.

- The private sector spent 0.66 % of the turn over on R&D as compared to 0.3 % by the public sector
- In industrial R&D (public and private sector) the share of expenditure by the major sectors in 2005-6 was as follows
  - Drugs and Pharmceuticals - 37.4 %
  - Transportation - 14.7 %
  - Defence Industries - 6.9 %
- Most of the central R&D agencies sponsor R&D projects on other institutions also and the comparative sponsorship data is as follows:-

Sponsoring agency	Value of sponsored projects-Rs. Crores	Number of sponsored projects
ST	572.11	1297
MCIS	183.85	42
DBT	174.73	422
ICMR	57.29	207
DRDO #	30.37	137
CSIR	24.21	247

It may be observed that the quantum as well as the number of projects sponsored by DRDO is very little as compared to other agencies.

· R&D efforts are usually compared internationally in terms of the % of GNP spent on R&D. The comparative data for the same is given below:-

Country	R&D expenditure as % of GNP
Israel	4.5
Sweden	3.8
Japan	3.4
Korea republic	3.2
USA	2.6
Singapore	2.4
UK	1.8
China	1.45
India	0.89

34. It is observed that the R&D expenditure in India is very low as comp ared to other countries including China. In India 3.91 lac persons are employed in R&D organizations and 39.6 % of that are directly engaged in R&D activities and the rest are in supporting roles.

- Patents are an important indicator of innovation. In 2006-7 totally 7539 patents were granted in India and 74.7 % were granted to foreign citizens and Indian share was only 25.3 % .The number of patents taken in the defence sector are very low.

(Source-Research and Development Statistics at a Glance2007-8 - Department of Science & Technology)

35. Research &development base is spread over the country in the following organizations:-

- Defence Research &Development Organization (DRDO)
- Ministry of Science & Technology (DST)
- Council of Scientific & Industrial Research (CSIR)
- Department of Space
- Department of Atomic Energy

- Ministry of Information & Communication Technology
- Academic institutions
- Private sector industry
- Public sector industry

36. Defence Research &Development Organization (DRDO) is the main organization engaged in R&D related to Defence.The Defence R&D expenditure has been roughly 6 % of the total Defence budget and less than 0.9% of the GDP. Research and Development for the defence sector has been done mainly by Defence Research & Development Organization (DRDO) in India. DRDO was established in 1958, with a limited mandate of advising Armed Forces and Defence Production units in systems analysis and sorting out design and development related problems. Most of DRDO establishments are co-located with Defence Production units. Initially, DRDO was engaged in the development of small arms and ammunitions. During 1970 – 80, it started work in reverse engineering and started gaining know – how of system engineering. During 1980's, thrust was given to major programmes like development of guided missiles, electronic warfare systems, radars, sonars, main battle tank, light combat aircraft etc.

37. Their mission is to design, develop and lead to production of state - of - the - art weapon systems, platforms and allied equipment etc. It aims at progressive enhancement of self - reliance in defence systems and make the country independent of foreign technologies in critical areas. It has a vast network of 50 laboratories spread across the country. It has a total manpower strength of about 28,600, which includes about 7,120 scientists from engineering, science and services, 12,750 technical staff and 8,730 administrative and other staff.

38. DRDO established Defence Institute of Advanced Technology at Pune which has attained status of Deemed University in 2005. It organizes courses on wide spectrum of technologies to meet defence requirements. The Institute of Technology Management at Mussoorie meets advanced managerial requirements for the scientists and other personnel. They also have training institute at Jodhpur to cater to the needs of administrative and allied cadre.

39. DRDO undertakes mission mode projects involving deliverables for the Services, technology development projects to develop demonstrators, science and technology projects for emerging technologies and infrastructure projects for setting up infrastructure facilities. To execute these projects, DRDO interacts with Defence PSUs, R & D laboratories, private

entrepreneurs etc. To complete the projects "Concurrent engineering" approach has been adopted in technology intensive projects to minimize time lag between development and production of the systems. DRDO involves users and production agencies from the very beginning to cut short the delays and bring synergy among developing agencies, R & D laboratories, users and production agencies.

40. Their success can be judged from the fact that the value of production orders placed on DRDO developed systems stands at over Rs 30,000 crores till date. This is out of a total R & D cost of about Rs 12,000 crore. Of this, approximately Rs 12,500 crore of production has been executed and the systems inducted.

41. **Programmes and Projects.** Their projects and programmes are broadly categorized into major disciplines like: missiles, aero-systems, electronic systems, combat vehicles, armaments, naval systems, advanced materials and life sciences.

42. **Technology Acquisition.** A new Directorate for Technology Acquisition has been created. It studies S & T Roadmaps of DRDO and the three services and identifies the critical technologies that are required to be established within the country. It will lead to strategies being evolved for establishing the technologies in the country by one of the following routes:

- In – house development in DRDO.
- Joint development through National S & T labs and academia.
- Joint Development though International Collaboration.
- Acquisition of Matured Technologies through offset obligations.

Integral to this objective, Director Technology Acquisition represents DRDO in the defence Offset Facilitation Agency (DOFA) of MoD in identifying technology areas where offset obligations can be successfully utilized. Some of the technologies and facilities identified by DRDO are given in Chapter 2 .

43. **Self – Reliance.** DRDO has till date developed a number of system and products, of which a large number have been productionised. The value of production orders placed on DRDO developed system stand at over Rs. 30,000 crore till date. Of this, approximately Rs. 12,500 crore of production has been executed and systems inducted. Considering the size of their set – up and the number of years for which they have been operating, this achievement cannot be considered satisfactory.

(Reference: Annual Report 2007-08, Ministry of Defence, Government of India.)

44. At present, R & D programmes of DRDO are, by and large, based on the requirement indicated by the armed forces. They, in turn, tend to relate the requirement to systems already in operation some – where in the world. Hence, R & D work starts with a bit of time lag, which is difficult to make up, and the net result is imports. Much of the R & D efort put in does not result in a usable product. At times, the product is available for use at the last stages of the life – cycle of the product. Hence, quantities produced do not result in full amortizing of the cost incurred in R & D effort, let alone making profit on the same. Needless to say that it demoralizes the R & D team, as well as, those who have put in ef fort to productionise the system. DRDO package doest not include production technology, which has also to be developed by the factories and it results in delay in commercializing the products.

45. A major change is required in approach to R & D.To start with, technology leadership in the world should, at least in a few products, be the objective of DRDO. In this way technology push will be provided by DRDO, driving armed forces to take on such products ahead of the rest of the world. This could also result in export of such products, which is required to achieve economies of scale in production.

46. At present, DRDO is not bringing out break – through technologies.This is clear from the fact that the number of patents obtained by DRDO are not in line with the size and age of the organization. Working on break – through technologies involves high cost and high risk. Success in this effort also leads to substantial profits. Given resources available in the country, break – through technologies can be brought out and they can also be commercialized. Israel has done it and China is in the process of doing so.

Interaction with DRDO

47. On the subject of self – reliance, it was mentioned to DRDO that against the target indigenous procurement of 70%, actual achievement is barely 30% in the country. They felt that to achieve the desired results, this objective should be that of the entire Ministry of Defence (MOD) and not of DRDO alone. Policies and procedures for procurement should be framed by the MOD to enable achievement of this objective. In fact, they should be binding on all concerned. Products developed by DRDO are of ten not accepted readily since stricter standards are followed for inspection of indigenously developed and produced equipment. At present, DRDO designs and develops products, and is responsible for taking them up to small series production. It will help in cutting down development cycle, if potential producers and users can be associated with the project at prototype stage itself. Over the years, however, development cycle is coming down though it may still be around 25% more than the international standards.

(Reference: Above comments are base on interaction with senior officers of DRDO)

48. The working of DRDO has been examined recently by the following two committees:-

1. Standing Committee on Defence (2006-07), Ministry of Defence, DRDO, Fourteenth Report.
2. Rama Rao Committee.

49. The Standing committee has made a number of recommendations.Those relevant to this study are as follows:-

- Private sector participation should be encouraged and relevant organizations should be given autonomy to perform. Weapon system should be developed concurrently by at least two corporations to encourage competition.
- R & D work on medical, life and other sciences should be entrusted directly to the concerned organizations. DRDO should concentrate on Defence and strategically important matters.
- A number of scientists have left DRDO in the last 10 years. In fact, 785 scientists have quit during the last 3 years. National policy should be formulated to recruit and train talented and experienced scientists who would make the pursuit of science a viable academic and commercial proposition.
- R & D budget should be 14-15% of the total defence budget against 4.87% in 2004
- From day one users and manufacturers should be involved from top policy making decision to all other levels
- DRDO must concentrate and focus on augmenting basic science and technological output to be at par with the other developed countries.
- Capabilities of IIT and other scientific universities should be utilized to build up defence capabilities.

50. Rama Rao Committee Report has given following recommendations that concern this study:-

- Emerging technologies that need to be focused on are: Biotech, and Aerospace design.
- There is requirement of large Public/ Private partnership.
- Critical sectors of science and technology are: Nano-technology , Material processing-fabrication, including electron and photon scales.

- Bio materials.
- Computers and Communication technologies.
- Armament technologies required are: Better Sensors, Faster Computers, Stand-off weapons, Enhanced transparency/ digitalization of Battlefield and Seamless communication, to mention a few.

Council of Scientific and Industrial Research (CSIR)

51. CSIR has multidisciplinary multi – location network of 38 national laboratories and 39 outreach centres. It has manpower of 17,432 with annual budget of the order of 1,710 Crores (2006-7).The CSIR was created in 1942 as an autonomous society In the first two decades after independence, CSIR built up an extensive infrastructure for R&D. It provided support to the emerging indigenous industry, mostly in small and medium sector. In seventies, it was assigned the task of developing indigenous technologies for SME's. It took up the same for a wide range of products.

52. In 2001, mission of CSIR was defined as "To provide scientific industrial research and development that maximizes the economic, environmental and social benefit for the people of India." The emphasis then was placed on proactive business driven R&D. CSIR today has near world – class facilities and infrastructure in most areas. It has emerged amongst the top three entities from the developing world in terms of PCT patent application filings and accounted for around 60% of all US patents granted to Indian inventors in 2002 – 03.CSIR has been ranked 8th in the world by WIPO for filing patent applications among the public funded R&D organizations in the world.

53. Areas where CSIR has proven strengths and potential include: genomics, biotechnology, Chemical Sciences and technology, new drug discovery, Clean leather processing technologies, and Aerospace engineering. It is strengthening its capabilities in emerging technology areas such as nutrigenomics, texigenomics, nanotechnologies, hydrogen and fuel cells etc.

54. For Defence Industrial Base (DIB), their capabilities in the following areas are relevant: Aerospace Science and Technology, Chemical Sciences and Technology, Electronics and Instrumentation, and Materials, Minerals, Metals and Manufacturing. They have various network projects on hand like: catering to specialized aerospace materials and High Science and technology for national aerospace programmes. The latter deals with generation of crucial design and certification data for aerospace systems.

55. Following laboratories, among others could be important for DIB:-

CSIR Laboratories in India

AMPRI	:	Advanced Materials and Processes Research Institute, Bhopal
CEERI	:	Central Electronics Engineering Research Institute, Pilani
CMERI	:	Central Mechanical Engineering Research Institute, Durgapur
CSIO	:	Central Scientific Instruments organization, Chandigarh
IICT	:	Indian Institute of Chemical Technology, Hyderabad
NAL	:	National Aerospace Laboratories, Bangalore
NCL	:	National Chemical Laboratory, Pune
NIO	:	National Institute of Oceanography, Goa
NML	:	National Metallurgical Laboratory, Jamshedpur
NPL	:	National Physical Laboratory, New – Delhi

56. Their network projects which could have relevance for DIB are: "Development of key technologies for photonics and Opto – electronics", "Developing cap abilities for micro-electro mechanical systems and sensors", and "Developing and sust aining high science and technology for national aerospace programmes".

57. For the purpose of study on the Defence Industrial Base, it is clear that CSIR has an important role to play if the goal of self – reliance is to be achieved by 2025. There are many areas of R&D wherein CSIR and DRDO can work together . It is p articularly so in dual use technologies for defence and civilian applications. They can work together on creating new knowledge, which is important, if India has to take a lead in technology front in the world.

58. CSIR have launched **New Millennium Indian T echnology Leadership Initiative (NMITLI)** in 2007 to catalyze innovation centered scientific and technological development, to attain for Indian industries a global leadership position, in selected niche areas. Key features of NMITLI are:-

- NMITLI synergizes competencies for Public funded R&D, Academia and Private sector.
- NMITLI projects are proactive and promote Low investment, High risk technology areas, with global leadership potential.

- Successfully funded 57 largely networked projects with outlay of Rs.500 crores involving 80 industry partners, 270 R&D groups and 1700 researchers.
- Application in following areas of Nano-materials and coatings, Fuel cell based on hydrogen, Next generation plasma display technology , Wireless sensor network chipset.
- Distributed video surveillance system.

59. The study of CSIR operations was done to examine its relevance to defence sector. In this connection, the following observations were noted :-

- Around 20% of the activities of CSIR are related to the strategic sector.
- They work on defence projects based on orders placed or them for R&D.They are not proactive on developing technologies, on their own, which may have possible application in the defence field. Of course, some of the technologies developed by them for use in civilian sector have application in defence sector too.
- There is a feeling that indigenously developed products are considered comparatively inferior by "defence", as compared to the imported products. Hence, there is reluctance on the part of indigenous industry to develop products for "defence".
- DRDO gives some project to CSIR. However, there is no regular interaction between the two large national R&D organizations. They do have overlapping of activities among them.

60. There is a scope for creating "new knowledge," if DRDO and CSIR work together and share their expertise and resources. CSIR could concentrate on civilian uses of the new knowledge and DRDO on the defence applications. This way, the country's preparedness for self – reliance by 2025 could receive a boost. New knowledge may relate to Biological warfare, Chemical warfare, Fuel cells, Bio – fuels, Long lasting Batteries etc. Both the organizations could aim at world leadership in some of the technologies.

- Commenting on longer time period involved in developing new products, the feed back was that it is largely a "Systems failure." Moreover, urgency is not noticed in handling R&D projects. It was felt that "Holding Company" structure may also not solve the problem of long period of time required for developing new products in India. Mainly because of bureaucratic way of functioning of enterprises in the Government/ Public sector. Involvement of private sector in this task may be helpful in achieving this objective.

- Development of new technologies at the initiative of CSIR/ DRDO could provide technology push to defence. Practical solution would be for defence to conceptualize new defence equipment. There – after the work could be allocated to DRDO/ CSIR/ Pvt Sector, individually or collectively.

61. Kelkar committee report on "Reinventing the CSIR", July 2004, listed a few concerns regarding functioning of the CSIR. Some of them are listed below:-

- Inability of the system to respond with requisite agility to emerging opportunities.
- Diminishing attractiveness of a scientific career in CSIR.
- Increasing culture of bureaucracy in the system.
- Inadequate appreciation and underst anding of technology implement ation and strategy.
- Inadequate skills in business processes and customer relationship management.
- Unavailability of skilled researchers in specialized niche domains.

They have also made a number of recommendations for enhancing ef fectiveness and improving efficiency of CSIR.

62. For the purpose of study on the Defence Industrial Base, it is clear that CSIR has an important role to play is the goal of self – reliance is to be achieved by 2025. There are many areas of R&D wherein CSIR and DRDO can work together . It is p articularly so in dual use technologies for defence and civilian applications. They can work together on creating new knowledge, which is important, if India has to take a lead a technology front in the world.

References:

- *"Reinventing CSIR", Kelkar Committee Report, July 19, 2004*
- *Interaction with DR. D. Yogeshwar Rao, Head, T echnology Networking and Business Development Division, CSIR.& Dr . Naresh Kumar ,Head of Research Development and Planning, CSIR.*
- *Annual Report of CSIR-2006-7*

Department of Science & Technology (DST)

63. DST's mandate includes formulation of related policies and co-ordination of areas of Science and Technology in which a number of institutions and departments have interests and capability. They support a large number of national institutions for basic and applied research. They also promote in house R&D in the private sector through various incentives. They work very closely with large number of institutions engaged in the area of basic sciences, research and technology development. Detailed information is available at the website of DST (www.dst.gov.in). It is suggested that DRDO /MOD should develop closer links with DST to identify institutions who can do research related to Defence R&D.

R&D in the private sector

64. The Department of Science and Technology have a scheme for the recognition of Research & Development for the in-house R&D in the private and public sector. The recognition is granted on the basis of adequate manpower, infrastructure and R&D activities. There are more than 1400 units recognized as R&D units in the country. These units spend approximately 0.58 % of their sales turnover. The R&D intensity is comparatively high in the following areas:-

- Drugs and pharmaceuticals 2.49 %
- Agricultural machinery 2.09 %
- Biotechnolog 1.81 %
- Earth moving machines 1.56 %
- Defence industries 1.52 %
- Telecom 1.44 %

There are five units in the private sector who have been recognized for R&D in Defence sector.

65. R&D in the Central sector - The Central Government is fully aware of the importance of Science and technology,; more than 70 % of total expenditure for R&D is done by the Central sector. Percentage of expenditure by objectives is as follows. It may be observed that Defence has the maximum share of 25.6 %. Percentage share of central sector R&D expenditure by objective, 2002-03 is given in the table below.

Objective	% of expenditure
Defence	25.6
Space	18.0
General Advancement of Knowledge	16.7
Development of Agriculture, Forestry and Fishing	12.4
Production Conservation and Distribution of Energy	7.7
Promotion of Industrial Development	6.3
Development of Health Services	3.6
Protection of Environment	2.6
Exploration and Assessment of Earths, Seas, Atmosphere etc.	2.3
Development of Transport & Communication	1.5
Other aims	3.3
Total	100.00

Manpower Status

How to increase the productivity of the knowledge worker is the challenges of the twentieth century.

– Peter Drucker

66. Human resource is the key driver to any enterprise undertaken. The best of technology and the best of equipment will have no productivity in the absence of qualified manpower to exploit the same.Today in the information age it is the knowledge worker which includes the worker who has the technical know-how who is the biggest asset of any organization. Drucker states that in the modern business the biggest asset is the knowledge that the worker has and which he applies for the productivity of the company. The world is facing a resource crunch of qualified workers and this has caused the mass exodus of workers trans-countries and continents. Thousand of Indian workers support the IT industry in the US and Europe. Rising competition and mass consumption have placed a premium on the price commanded by the knowledge worker and he moves where the work conditions and compensation is better.

67. While the Private sector also have their share of problem in recruiting staff, they enjoy an advantage of fast decision making in tune with the market forces of demand and supply

unlike the private sector. The human resources that make a significant ef fect are at the working level and at the top management level.At both level the level of migration is high. The current turnover level of the knowledge worker is as high as 45 %. This poses a serious challenge to the success of the Industrial base in generally and to the PSU' s in particular.

68. The genesis of the manpower lies in the policy post independence where the PSU were expected to att ain the commanding height s of the economy . These organizations were run by the government as an agent of economic and social change. Therefore the influence and control of the government was great. Most of the top management was from the civil services this naturally extended the government work culture to manufacturing units. The most damaging was the feeling of permanent employment backed by strong Unions. The pay scales were more in line with the government services rather than at the market rates. The exodus from the public sector to the private sector whenever the opportunity presented was but natural. Post 1991 as the economy opened up and more jobs were created there was a short age in the private sector . This led to a move of the experienced personnel to the private sector . The boom in the IT sector caused a move from India to the countries of the west paying in dollars.

69. Intake of personnel at the induction stage is from the colleges and technical institutes. The private sector manages to mop up the cream of the talent with compensation packages at par with the market and at times that have no relation with their work experience. This leaves the PSU's to get recruit s from the second rung colleges or private institutes with dubious st andards. Even these recruit s shif t to better p astures once they gain some experience.

Manpower in DRDO

70. As per the report of the Parliamentary Standing committee on Defence 2006-07 of the 14th Lok Sabha, the DRDO reported to the following to the committee:-

- Total authorized strength of manpower - 33442
- Total Scientist who left in last 10 years - 1404
- Reasons for leaving were Personal/ Domestic.
- DRDO is a Project based Organization and no project has suf fered for want of manpower.

- Salary structure as per 5th pay commission permits two additional increments on joining to all scientists up to the slab ofRs 16000 – 20000, thereafter a special pay of Rs 2000/- per month for scientists in the slab of Rs 18400 – 22400.
- A bonus of Rs 5000/- per month per annum for all scientists after peer review.

71. An analysis of the above reveals that these incentives in the p ay structure were enjoyed since 1 Jan 1996. The comparative increase in salaries in the corporate world started only around 2004 and as such was at par with the corporate world and the best in government. This is adequately borne out by the fact that 1404 scientist lef in a period of 10 years. DRDO has not given a breakup of this figure and it is not hard to visualize that it would be in the later p art. The average percent age works out to be 4%, which is not admirable as per the corporate figures but simply brilliant. Therefore to assume there is a brain drain would be incorrect. This very low turnover can only be assumed to be for two reasons either the work force is happy or their quality of performance does not ofer them opportunity elsewhere. Therefore to attribute non performance to this factor would be wrong.

72. The study team hastens to add that it strongly recommends recruitment and retention of talent but this talent has to be defined. The bonus of Rs 5000/- per annum on peer appraisal has no organizational benefit except to stick together so that each scientist gets this, year end bonus. It is doubted if any one would be left out.

73. The parliamentary committee has recommended additional incentives as a measure to retain talent. While the wisdom of a committee is not questioned certain comments can be offered these are offered in three parts first where it motivates a person to perform and second where it is desirable as part of the developing social structure and hygiene factor. The third is a wish list which if achieved is great but is of no consequence.

Motivating Factors

74. The motivating factors are as follows:-

- Enhancement of professional update allowance.
- Authorization of additional increments to Scientists 'B' at the time of joining.
- Enhancement of Study Leaves for acquiring higher qualification namely PhD, M Tech, etc. or Grant of Sabbatical Leave for interaction with academia and R&D departments.
- Detailing scientists for Short Term/Long Term training courses in frontier areas within the country and abroad.

- Fast track promotion policies.
- Sharing of monetary benefits of commercialization of products/processes developed by inventors.

75. The best incentive is that at serial 6 above which is in line with the Organizational and Individual interest. Study leave or sabbatical must be available only once unless there are extenuating circumstances that the same person has to be trained for the said discipline to be approved by the head of the DRDO personally. All leave and courses are subject to a bond that the individual will not leave service for three years.

Social & Hygiene Factors

76. The factors are as follows:-

- Internet facility to all the scientists at home.
- Award Scheme to be augmented.
- Augmentation of residential facilities, recreational, medical and educational facilities.

77. As the economy grows there must be a correspondence of spreading the advantages of higher earnings to all rungs of society. For government servants these can be offered only in kind so that permanent assets are built up and also act as a motivating factor. Better social life can not be created by any organization but they can provide common facilities for the workers and their families which can contribute to build it.

Wish list

78. The following fall in the category of a wish list:-

- Individual Consultancy & Collaborative opportunities.
- Entitling all scientists for Air Travels on Official/Temporary duty.
- Reimbursement of residential telephone and Internet expenses for young scientist also.
- Adjunct appointment.
- Better social life in the DRDO living campus.

79. These factors are governed by rules common to all government employees. Individual consultancy can only be at the government expense and can be clubbed with the sharing of monetary benefits as motivators. Social life depends upon the individual himself and

has been covered at serial 3 of the social and hygiene factors. These not only enhance transparency and p articipative behaviour but also stimulate the rank and file towards performance and accountability. A quote by Shri A S Seetharamu in Business Line, The Hindu 14 Mar 2004 is as under:-

> *"Setting, seeking and self-realizing a vision is imperative for every PSU. This is the first step in PSU reform and should form the core of HRD programmes in simple, clear, intelligible and coherent language. Sharing of vision with the workers effect, they contribute to the appreciation of productivity , efficiency and competitiveness of the firm ".*

Training of Manpower

80. Training is about building and upgrading the technical competence of the manpower in the country to sustain the economy and industry. In the public sector, the upgrading of the competence of the personnel is being carried out by in-house training and for the senior personnel from premier colleges of the country . The administrative st aff college Hyderabad conducts training senior officials.

81. The Justice Mohan committee constituted in 1996 recommended that the playing field between the private sector and the public sector is not the same. Further the dif ference between the lowest and highest salary in a PSU in the ratio of 1:10 is considered to be adequate. Also the pay structure in a PSU should be based upon the financial position and earning of the unit.

82. The lack of adequate competence in the Public sector units can best be summarized in the words of the late Prime minister Charan singh when he laments that, *"In the Public sector undertakings of our country , the situation has, inter alia, been bedeviled by the patronage dispensed arbitrarily by political leaders and their blat ant efforts at playing to the gallery and pushing up to top management level the so-called 'committed' elite devoid of business acumen and requisite managerial skill who enjoy their assignments 'smug as a bug in a rug', secure in the knowledge that they would not be held account able for the losses".* (Choudhary Charan Singh in 'India' s economic nightmare – National publishing house 1981').

83. More recently the Prime minister has set up a National Knowledge Commission (NKC) on 13 June 2005 to assess the development of a knowledge based environment required in support of a knowledge based economy and Industry in the future. The Prime minister stated *"The time has come to create a second wave of institution building and of excellence in the field of education, research and capability building so that we are better*

prepared for the 21st century." The terms of reference and objectives of the commission are given at given below.

National Knowledge Commission

Terms of Reference

As per Government Notification of 13th June 2005, the following are the Terms of Reference of the National Knowledge Commission (NKC).

- Build excellence in the educational system to meet the knowledge challenges of the 21st century and increase India's competitive advantage in fields of knowledge.
- Promote creation of knowledge in S&T laboratories.
- Improve the management of institutions engaged in intellectual property rights.
- Promote knowledge applications in agriculture and industry.

Promote the use of knowledge capabilities in making government an effective, transparent and accountable service provider to the citizen and promote widespread sharing of knowledge to maximize public benefit.

Objectives

The overarching aim of the National Knowledge Commission is to enable the development of a vibrant knowledge based society. This entails both a radical improvement in existing systems of knowledge, and creating avenues for generating new forms of knowledge. Greater participation and more equitable access to knowledge across all sections of society are of vital importance in achieving these goals.

In view of the above, the NKC seeks to develop appropriate institutional frameworks to:-

- Strengthen the education system, promote domestic research and innovation, facilitate knowledge application in sectors like health, agriculture, and industry.
- Leverage information and communication technologies to enhance governance and improve connectivity.

Devise mechanisms for exchange and interaction between knowledge systems in the global arena.

84. Historically, India has had an enviable standing in the world of research. However, the current situation is unsatisfactory across multiple dimensions, and the reasons for this are

manifold. An important factor which imp acts the quality and quantum of research is the scarcity of t alented and dedicated manpower engaged in research. W e need qualified doctorates for our research laboratories, to teach and engage in research in our universities and to work for research and development in all sectors of the industry . The number of researchers in India was 112 per million inhabitants compared to 633 in China and 4374 in USA in 2002.The growth in the number of doctorates has been only a modest 20% in India from 1991-2001 compared to 85% in China during the same period. The current state of affairs thus requires urgent policy intervention[1].

(Report of the NCW to the prime Minister. 1 "Measures of Progress of Science in India", Report by NISTADS 2006 – http://psa.gov.in/writereaddata/11913286541_MSI.pdf)

85. While a knowledge based environment is created it will have a direct imp act on the DIB. The NKC has made 27 recommendations to the government on various issues leading to world class education and a knowledge based society in the coming years. The parts most relevant to the DIB are:-

- Education at the doctorate level.This recommendation leads to the higher education and creation of a manpower force capable of research in S&T, to which the DIB is closely related. In the coming years, as the defence equipment grows sophisticated in technology it would need educated personnel at the appropriate level to man it and develop it. At the current rate there is likely to be a vacuum in the future and hence the urgency of motivating people to choose R&D as a career.
- Intellectual property right s (IPR). Creation of new or cutting edge technology , innovating new processes and methods of manufacture and employment technology imparts a superior edge, this is soon lost unless the IPR is safeguarded. Positioning laws and infrastructure to protect IPR is a as important as developing technology.

86. The commission has made relevant recommendations like revamping the education system at all levels and also a close interaction with the industry so that the education system is in sync with the needs of the industry . This will not only boost industry but will create job guarantee.

SWOT Analysis of DIB

87. The SWOT analysis of the existing DIB identifies the strength, weakness, opportunities and threats is given below:-

- **Strength**
  - Large captive market
  - Established infrastructure in Ordnance factories and defence PSUs
  - Experienced manpower
  - Large R&D base IN DRDO, DST, CSIR, ISRO, DAE
  - Well established private industrial sector
  - A growing economy
- **Weakness**
  - Policy regime in India is extremely encumbered
  - Technology used for production is comparatively outdated
  - Heavy dependence on import of complete equipment-70 %
  - Imported technology not absorbed / upgraded for higher version of products in most of the cases
  - Production operations are not competitive in terms of quality and delivery
  - R &D not innovative, only 66 patents over 30 years with large network of 50 labs and 30,000 manpower
  - Very difficult to identify areas where we are world leaders
  - Science and technology not attracting the best talent in the country
  - Number of PhD scholars in S&T is low and still lower in areas related to Defence R &D
  - contribution of private sector in defence production & R&D miniscule
  - Failures in R&D not employed as a learning process
  - Product development cycle time is extremely long, by the time product is developed it looses relevance and import takes place which gives further set back to R&D
  - Interaction with Academic world has started but it is too little

- Difficult to attract and retain expert technical manpower in Defence R&D and production system
- Indigenous Machine tool building capability has gone down
- Exports of defence products are negligible in value and with low value addition
- Direction for use of Off sets is not clear, though the organization structure has been set up
- Vested interest in maintaining status- quo especially for imports.
- Monopoly of defence production by PSU's difficult to break into.
- Recommendations by several high power committees including Parliamentary bodies not implemented due to Bureaucratic stranglehold.
- Academics work under a different work culture and are beast when given freedom of action

• **Opportunities**

- Economy is on the growth path
- Policies liberalized for participation of the private sector in DPP-2008
- Offsets policy is an opportunity for building DIB
- Industrial Licensing policy liberalized for Private sector participation in defence industry
- FDI up to 26 % permitted in defence industrial sector to facilitate transfer of technology.
- Private sector has taken Industrial Licenses for the manufacture of Defence products
- Strong private sector production base in the country
- large market in India as well as potential exports
- Technology imports earlier denied to Indian Defence industry likely to be eased.
- Large technical talent & companies available which could be tapped.

- **Threats**

– Vested interest in maintaining status- quo

– Bureaucratic stranglehold may not permit implementation of recommendations to reorganize Production and research.

– Difficult to let go the monopoly of defence production by Defence units

– Very Strong import lobby led by Dealers & Politicians

– Difficult to attract and retain talent in Defence R&D

– Academics are not used to challenging and time bound defence projects and they want more freedom

– Our adversaries are building strong DIB with modern technologies

– Lack of level field perceived by the private sector

88. **Reasons for Stagnation of DIB in India.**

- Excessive bureaucratic control
- Strong lobby for import
- Lack of competition – monopolistic conditions
- Poor indigenous R&D
- Lack of encouragement to the private sector
- Production system developed on Soviet model of Government control in Defence R&D and production with no regards to cost. This needs modification.
- Defence R&D spread thinly over many areas with little innovative output
- Technology development cycle from lab to field is very long due to lack of proper project management & absence of concurrent engineering

89. **Need for Change.**

- Growing economy needs self reliance in critical areas of defence
- Existing DIB is not geared to meet the need which has resulted in import of 70 % of our defence requirements. This has the following adverse impact :-

- Dependence on foreign countries for national defence
- Lack of national pride
- Out go of more than Rs.2, 250,000 crores in the next 15 years in foreign exchange (@Rs. 150,000 crores per year) if the existing situation continues. This would be at the cost of local development
- Creating jobs abroad and not in India.
- Recurring Imports demoralizes the indigenous R&D community
- Source of major scams which delay the modernization process e.g. no modernization in artillery since 1988.
- Existing DIB producing comparatively outdated products

- ♦ Economically desirable to make the production and R&D competitive in national interest.
- ♦ Imported technology will be invariably out dated.
- ♦ Leadership in few areas will regain national pride and give negotiating strength for acquiring / developing other technologies.
- ♦ Conserve foreign exchange.
- ♦ Neighbourhood of India unstable – need to modernize the defence forces.

Recommendations

Production

90. The recommendations are as follows:-

- Encourage private sector production by providing level playing field and through suitable mechanisms like development order for Make category of items.
- Restructure the ordnance factories to make them competitive through the following:-
  - Give them more autonomy by converting them into corporations.
  - Carry out BPR (Business Process Engineering) to rationalize their product profile.
  - Restructuring may be done through setting up holding comp anies having production, design and R&D under one CEO.

- Appropriate % of the cap acity may be reserved for strategic product s which have to be manufactured in-house by MOD. Balance capacity may be used for non strategic / other uses which can be diverted in emergency to increase the production of strategic products.
- Introduce the concept of competition wherever possible among the units in the government sector also.
- Make the organizations lean for high performance.Their turnover to manpower ratio is very poor when compared to any other leading.
- manufacturers in the SIPRI top 100 companies.

91. Change the mandate of Ordnance Factories to include:-

- Product development through in-house R&D (surprisingly, it is almost absent in the existing system)
- Technology absorption with clear understanding that mark-2 of the product has to be developed in-house. The process of technology absorption has to st art from day-1 and should be backed by suitable funding and manpower.
- Ordnance factories should spend 5 % of their turnover on R &D.
- Enhance R&D in the PSUs to a level of 5% of their turnover for product development as well as technology absorption.They should also have mechanism for sponsoring projects on other institution in public and private sector including the academics.
- The DGOF have been outsourcing supplies since 1970 but this has not resulted in building a second tier of suppliers & manufacturers due to Industrial policy until 1991. Policy Decisions therefore must be taken holistic manner.
- The DGOF must come out of the Government straight jacket and function as a corporation with freedom to develop product s and processes, explore & exploit markets and develop HR policy commensurate to it s needs and cap ability. The concept that investment in DGOF by the government solely for a single customer must be jetissoned.
- While approvals to private sector have been granted to produce a large range of products their actual production has yet to commence in fact even the production facilities have yet to be established. Therefore the situation is not optimistic.

- There is an imperative requirement to establish a coordinating agency at the apex between industry, defence and ministries.
- Although a slew of new measures have been t aken , the major bottleneck of bureaucratic control has not been diluted or removed. It is feared this may result in business as usual resulting in cosmetic change only sine subst antial results are yet to be seen.
- All research facilities in the country must be included to exploited to build the DIB.

Human Resource

- While the DRDO seeks ideal work conditions and facilities it does not promise performance in return. Therefore incentives must be conditional on performance.
- Study leave of any kind must be based on demonstrated cap ability in the past so as to be a further build up on the same. Also leave has to be concurrent with a commitment to benefit the organization and therefore need for a bond to serve four years further in respect of the scientific community. .
- Work and living conditions must be at par with the best and in consonance with the income of the government as part of hygiene factor and also a national commitment of sharing the benefits with all strata of the country.
- Other proposals under the wish list are subject to the rules applicable to all government employees.
- Internal communication within the campus must be made available to maximum possible personnel.
- Academic Institutions to be incorporated in the development of military technology as a Budgeted function by the Ministry of Defence.
- The recommendations of the NKC relevant to the DIB must be followed.

Chapter 5

Government Policies, Budgeting & Procurement

Policy

1. The dynamics of security are extremely complex, unpredictable, and operate simultaneously at multilevel. It requires a highly attuned and sensitive organization to respond to the emerging challenges. Security related challenges can only be ignored at own risk. In order to build an organization that is foolproof would be extremely expensive and not cost effective in fact it can be compared to a Pyrrhic victory. Traditionally the defense forces have been organized and equipped to meet all major threats that they might face. This dictated the organization, equipment and operational doctrine. With increasing technological warfare, fast communications and transport the capacity to meet all threats has caused an exponential increase in defense expenditure, leading to a thought process of developing a capability base force that is highly mobile well equipped and which can meet any eventuality due to its inbuilt capability of reach and effectiveness. As with all thought processes, opinion swings from one side to another as events and actions unfold. These events validate one theory or the other causing debate on which is the best.

2. The decision on developing a threat based vs capability based force depends upon the national situation, resources and the likely roles that our forces might be called upon to undertake. It therefore has to be an optimum solution taking into account all factors that favour national defense. Further decision must not get swayed or influenced unduly by experience of others and start to question own national aims. In the Indian context there is a definite requirement for a minimum force to meet the obvious threat based needs and an enhanced capability to meet the out of area contingency synonymous with our emerging status in the region and geo-strategic imperatives.To amplify Chinese have an ambition to deploy their navy in the Indian ocean, they have embarked on the strategy of *a string of pearls,* and are developing assets in Mynamar. An overland gas pipeline from Mynamar to china as also from central Asia to china is on the cards. In theory they can prospect for oil in the Bay of Bengal off Mynamar being international waters. These are factors that affect

India and it must build and demonstrate its capability to neuter these should the occasion arise by building and demonstrating the necessary cap ability. In the absence of a strong national DIB this may not be possible and with dependence of more than 70% on imports it will not be permitted. *Thus a national policy on security and defense is imperative to sustain long term military plans and their execution for the defense forces.*

Industrial Licensing Policy and Export Policy

3. Historically the production of defence industry was reserved for manufacture by the state Government as per the Industrial licensing policy of the country.

4. The private sector contributes approximately 10% of the needs of the defence services at two levels. The first is by the corporate sector consisting of a handful of companies who have an establish business and contribute the same or similar products to the defence. The small and Medium enterprises (SME) contribute more than 41% of the needs of the defence sector.

5. The private industry was not allowed to manufacture Defence equipment as the same was reserved for manufacture by the Government as per the industrial policy of 1956. This policy was framed by Prof PC Mahalanobis appointed by Nehru. He was assisted by three planning experts deputed by Moscow who were deeply influenced by the conditions in their country. Russia was surrounded by enemies and their primary concern was security and defence and all defence industry was placed in the public sector. This was revised in 2001. Some units in the private sector supplied some of the component s and fabricated items for defence units .In 2001 the Industrial Licensing policy liberalized to permit private sector with FDI up to 26 %. Till 2007 73 letters of intent s have been issued to various Indian Companies. The list includes some of the well established organizations like BHEL, Tatas, Mahindra & Mahindra, Larsen & Toubro, Samtel.

6. The policy was liberalized significantly in 1991 but the reservation of Defence industries continued. In the year 2001, defence industry was opened to the private sector. As a result of that more than 140 Letter of intents have been issued to the private sector.

7. As per the current Policy the production of all defence equipments require an Industrial Licence. The direct foreign investment in defence industry is restricted to 26 % as per the current policy.

8. The guidelines for licencing production of arms and ammunitions are given at **Appendix B.**

9. Export of defence items come under export control and export licence is required for the same. There are territorial restrictions on export also. The exports from India have been low.

S&T Policy

10. The Department of Science &Technology have a comprehensive policy on Science & Technology and it covers the objective and the issues related to basic Research, applied research and technology.

11. The Department of Scientific and Industrial Research (DSIR) has many programmes for giving Fiscal incentives for the promotion of Research and Development. (www.dsir.gov.in/tpdup/irdpp) Details of the same are given in **Appendix C.**

Defence Policy

12. Defence Industrial Base is governed by number of policies. The major policies relate to Defence Planning and budgeting, Defence purchase, Industrial licensing, export and Science & Technology. The relevant policies have been examined in this chapter.

13. Policy is the fountain head of all organizations and is the guiding light for all actions across the length and breadth of the organization. Today it is in vogue to condense this into a mission statement this acts as a constant reminder for all stake holders for the path to follow. It makes sound sense to align all actions to meet the higher aim of the organization. National policy on Security forms the basis of the defence policy to guide the armed force to work in unison to meet the national objectives. Policy planning in the context of the defence services has certain unique features that have to be catered to. The policy is contingent on the international environment, attitude of the immediate surrounding countries in the region and the anticipated threat perception. It also has to configure the available and emerging technology. Above all the process of planning, procurement, induction and exploitation is long drawn extending from 10 to 15 years. It needs to be reviewed regularly like a roll over plan to keep it updated in consonance with developments at home and abroad. With this as a background it is logical that the defence fiscal planning too has to be aligned with the Defense long term planning.

14. It is valid to peek into history to understand the present situation clearly. Before Independence the defence budgeting was done by the British government to achieve the military aims of their empire. Post independence this changed but regrettably no policy was formulated to spell out our national interests. The political leadership adopted a policy of non-violence and non expansionist ignoring the practical side of realpolitiks.The armed

forces had little or no say their leadership consisting of junior officers propelled into senior positions having no experience or strategic vision. Further the political leaders in a newly independent country still viewed the armed forces as a tool of British imperialism to be humbled at the earliest. They also had fears, emanating from what had happened in other countries that the army may seize power These conditions were hardly conducive to start a debate on formulating a national policy on defence. The first prime minister took the initiative to reduce threats by initiating the principles of Panchsheel in foreign affairs. He was not naïve but an astute educated man with understanding of national affairs who possibly tried to reduce defence expenditure by instituting a suitable foreign policy. He was lulled into this by China which had its own ambitions of a regional power; it was this that prompted it to invade India in 1962.

15. India has not articulated its national security policy and subsequent the Defence policy, this is a serious planning void and reason for not having a long term defence fiscal policy. The defence services as a professional organization have formed in-house long term perspective planning extending to 15 years, further broken down into 5 years plans which coincide with the national 5 year plans. This is the basis of Long Term Integrated Perspective Planning (LTIPP) by HQ IDS. This is with a view to have a rational, comprehensive and continuous plan to impart a sense of direction and make projections for budgetary allocations. It also enables the services to organize, equip and train the forces as per the master plan. To be Practical, all plans are required to have a budgetary support which is not the case and therefore in the absence of such fiscal support these plans are only indicative and subject to be dropped, changed or totally ignored at any stage making the entire exercise null.The closest that the government came to stating the policy at the apex level was the Defence minister's statement on 16 May 1995 given below-

Defence Minister on Defence Policy

16. The defence minister stated in Parliament on 16 May 1995, "We do not have a document called India's National Defence Policy But we have got several guidelines which are strictly followed and observed and those can be summed up as follows:-

(a) Defence of National territory over land, sea and air encompassing amongst others the inviolability of land borders, island territories, offshore assets and the maritime trade routes.

(b) To secure an internal environment whereby the Indian Nation state is insured against any threats to its unity or progress on the basis of religion, language, ethnicity or socio economic dissonance.

(c) To be able to exercise a degree of influence over the nations in the immediate neighbourhood to promote harmonious relationship in tune with the national interests.

(d) To be able to effectively contribute towards regional and international stability and to possess an ef fective out-of-the country contingency capability to prevent destabilization of the small nations in our immediate neighbourhood that could have adverse security implications for us.

17. This Policy is not merely rigid in the sense that it has been written down, but these are the guidelines, these are the objectives, these are the matters, which are always kept in view while conducting Defence policy I think no more explanation or elaboration is needed than this".

** From the above it can be safely inferred, that India, will zealously guard its interests within the country as well as in areas cont agious to it which have implications on it s security. It has st ated the obvious without revealing further details. It also has an element of flexibility, to react in an appropriate manner should an event occur in future which has not been specifically mentioned. Broadly speaking it gives subst antial guidelines for the milit ary machine to form their perspective plans and cater for the resources in order to build a reasonable capability.*

18. This however does not have the legislative or financial backing and is therefore subject to individual interpretations. This greatly affects the industrial thinking and planning on defence related investment in future. It clear articulation will also assist in military product development by the private sector.

19. While the Service headquarters form their own perspective plans being a professional force, these do not have the sanctity or approval of a national policy and fall in the category of a Wish-list and more often than not is referred to as such. It also pits one service against another for a larger share of the Budget a condition eminently ripe for bureaucratic patronage. People having no idea of the technical functioning of the Forces and their requirement sit in judgment over import ant issues (recall the case where snow scooters were required in Siachen). The Budget projections are therefore subject to pruning or cuts at the mercy of anyone between the initiator of Budget claim and it s final approval. This not only creates a situation of uncertainty abhorred by the Industry who would not like to risk investments on intransience approach.

20. In the absence of budget planning to support the services long term plans allocation of funds is carried out on an annual basis. Projection for funds is submitted by the services

based on their LTIPP considerations to the ministry of finance through the Defence ministry The finance ministry, generally, allocate funds based on previous years spending plus ten percent, in fact the defence budget is almost pre-decided. This is the mismatch between the two ministries and until resolved will continue to be ***the bottle neck*** affecting defence production and preparedness. Since there is no long term planning each acquisition is viewed in isolation leading to prolonged debate which may or may not result in the fructification of the project but certainly affects defence capability. It has been experienced that at times at the eleventh hour the complete process is negated and has to start de-novo. This also leads to supremacy of generalist in the decision making process ignoring specialist advice in the vital sector of national defence, where during war or conflict there are no runners-up.

21. The reason for this state is not difficult to fathom. Long term defense planning means a financial commitment for the plan period. Since defence takes away a large portion of the budget this cannot be decided in isolation as it has an impact on the national development, which means that the planning commission has to be consulted. Long term plans therefore have to be vetted and factored in by the planning commission the fountainhead which forms the source upon which the finance ministry evaluates the budget estimates received. Secondly defense has been a holy cow which has never been questioned even in Parliament and most starred questions have been stonewalled under 'secrecy due to reasons of national security'. This led to creation of a close knit power center which has the tacit approval of the finance ministry. The situation has changed with growing awareness, power of Internet and the voice of the Media. Citizens want greater clarity and wish to know what the government is doing for national defence, failures are being questioned and casualties of soldiers are unacceptable. The solution therefore has to be worked out between the defence ministry, finance ministry and the planning commission.

22. Post Kargil and formation of Integrated Defence Staff (HQ IDS), the responsibility of coordinating and monitoring of the defence budget has been entrusted to it. At present this role of HQ IDS is perfunctory only due to lack of support staff and other where with alls including appointment of a CDS. As part of a dynamic ministry of defence, HQ IDS needs to be incorporated more actively in the planning process to look after the professional and technical aspects. They also have to be incorporated in the formulation of long term defence financial planning. Of course there will be resistance to change (tremendous) as this will enlarge the decision making machinery and thereby accrue greater transparency and accountability. The major advantage would be that financial allocations will flow almost automatically in line with the pre-decided defence policy and spending to implement the

same. It will reduce ad-hocism and the whims of the current government in power to negate, dilute or divert funds to the priority areas of the Party at the cost of national defence. This is the reason why there is procrastination on appointing of the CDS under the plea of wider political acceptance. The bureaucracy is actively abetting, by projecting it as a matter of civil control over the army. No political party would subscribe to their dilution of political or fiscal control. The truth of the matter is that political control is vested in the Parliament and the supreme armed forces commander the President of India, and not in the bureaucracy At present the fiscal control is not even with the defence ministry and continues to rest with the finance ministry. The Defence ministry is only carrying out the expenditure of the allocated funds.

23. We are today facing a crisis of direction where there is ambiguity at the highest level causing confusion not only in the armed forces but also in the bureaucracy, industry and the scientific and research community. The result of this ambiguity was the Chinese aggression in 1962, and the subsequent wars as well as the present fears that the defence forces are not be in a position to perform their role effectively. Projections of non existent fears for turf protection by certain quarters are totally out of context and are seriously affecting national defence capability.

Budgeting

24. Defence budgeting must be examined in a holistic manner and it is not similar to that of other economic and social sectors. While the objectives of the social sector can be fixed and costed to a fair degree of accuracy, the same does not hold true for defence. To begin with the threats to the national security can only be an educated guess and the resources required are worked out based on this threat perception. But can the threat that will actually emerge be anticipated? *History has proven the converse to be true, in fact successful offensive operations in the history of warfare have been those where the planners have been unconventional and gone against all accepted concepts*. This clearly and firmly drive home the fact that our forces should be organized based on their capability and not based on threat perceptions. Further, recurring expenditure to maintain a force requires revenue expenditure. This by itself is different under peace conditions but its level cannot be predicted during war. The price of capital equipment and technology also is uncertain and depends upon the seller, the purchase and most importantly the mutual equation. This mutual equation is a fleeting relationship that includes politics, self interest, economics and exploitation of the situation and can therefore be extremely complex (the Air craft carrier Groshkov case). In order to ensure a planned, logical and systematic execution of Budget the following sequence is desirable in that order.

- Defence policy to be spelt out.
- Assess the threat perception
- Work out the objectives
- Evaluate resources required to meet the objective.
- Cost the objective as individual program.
- Plan the expenditure & budget for the same as a program.
- Fix responsibility to meet the program objectives
- Monitor the expenditure
- Annual audit by CAG to be tabled in parliament.

In order that these issues are successful for a streamlined process they are discussed as follows:-

25. **Policy Direction** - The defence policy flows from the national security policy. The aim of policy is to identify the defence needs and how to implement the same by setting tangible Objectives. Objectives in a milit ary organization are achieved by equipping the force with the weapon and equipment and to train and sust ain these resources. The maintenance of the forces forms the revenue expenditure and not part of this study . The capital expenditure of the Budget caters for modernization and fresh acquisition. Acquisition, in any armed forces are inducted over a period of time for full exploitation. Generally there are three generation in service concurrently. One generation that is being phased out, the main generation in use and the new generation under induction into service. This has been reiterated by the Arun Singh committee report. The capital expenditure is therefore generally spread over a few years. If this is well planned a fair estimate of expected expenditure can be made. More ever the major hurdle of acceptance of necessity and technical evaluation is already over and not subject to ever ending debate. This will also facilitate the ministry of finance to question the implement ation of the project as it s right and also permit the government to raise or lower the priority of the Project depending upon other national imperatives and urgency.

26. **Threat Perception** -Based on the national policy on security the defence ministry will examine the area of responsibility. This will include the following:-

Primary Capabilities

- Secure the national frontiers of the country.
- The capability to meet the need for internal security & aid to civil power that may be required of the services.
- Protecting the coastline and securing the EEZ
- Securing the national Air-space from any intrusion.
- Develop capability of the Air Force to provide close support to the land forces.

Strategic Capabilities

- Develop a credible nuclear deterrence.
- Develop Net work centric warfare (NCW) capability.
- Dominate the Air space over the region of Interests and develop capability in outer space.
- Secure the homeland from an attack by strategic missiles and artillery.
- Develop capability to intercede in the land mass contiguous to the country and in the area of interest.
- Dominate the sea routes of commerce and trade of the country.

Once these areas of national concern have been decided the objectives can be worked out.

27. **Setting of Objectives -** The defence objectives can be worked out specifically in respect of each Service that would be required to be attained to achieve the defence of the country and its interests. It is well understood that these objective must be capability based rather than threat based. The discussion on deciding this would be broader based after including the views of the other ministries whose area of interest s it would be the responsibility of the ministry of defence to safeguard. These objectives would therefore emerge as a broad consensus at the national level. Specific to the services these objectives would be defined as projects that will have inbuilt into them, what is to be achieved, in what time frame, what are the resources required and how these will be made availableAbove all who will be responsible for the implement ation of the project. (In the corporate world incentives and penalties are also spelt out).

28. **Resources required to meet the objectives** -Having decided upon the objectives the ministry of defence in agreement with the Service headquarters can address the specifics of the resources that will be required to meet the objectives in terms of military hardware. Here most importantly it will be necessary to evaluate the numbers required and also the technology to be utilized. High technology need not necessarily mean reduction of the force level but developing the capability of meeting the objectives in a more precise and efficient manner. It would also mean that the national capability is not at a disadvantage vis-à-vis the major adversaries. Defence technology is in a constant state of flux in an attempt to gain an edge over others. This necessitates that our policy should have inbuilt into it the method to constantly upgrade our equipment and to replace it periodically. This ensures clarity of purpose in the long term planning and catering for the funds. Else it gets bogged down in ad-hocism and each project of modernization gets bogged down in a maze of debates, pressure and views of vested groups. The focus of equipping the forces gets shifted to that of the purchase transaction and result in reducing operational capability and defence un-preparedness. It needs no examples to illustrate this since all our major defence procurement have followed this undesirable route. Further it also results in ambiguity and loss of a sense of direction insofar as the defence R & D efforts are concerned. It also affects the military doctrine so essential for military planning and training.

29. **Costing of the objectives** -Once the process of identifying the resources required is known the next question is what is the quantum of funds required? Also whether it would be economical to buy or make them. Funding has two parts the first is the capital expenditure to buy the equipment and the Revenue expenditure that is reoccurring each year to maintain this equipment. This would include the manpower, storage facility, maintenance spares, FOL and ammunition in respect of weaponry. Costing in the defence industry is the most fluid factor and in each case the cost is likely to be different. It depends upon the political equation between the two countries, their economic state and the state of mutual trade and commerce. Basically a seller would like to exploit the best he can for his years of efforts to develop the item. Conversely the buyer would like to leverage his purchasing power and reduce his vulnerability in negotiations by other means. For instance on the issue of transfer of technology the cost of technology can never be precise and a buyer can ask for any cost depending upon the urgency of need by the buyer.

30. **Budget planning to meet the cost** -The budget planning in case of the defence services is a little different from that of the social & economic sector. In pure economic terms defence budget is a non productive investment with no returns. It is a necessity that a country has to live with for its security. The attempt is always to achieve the objective at minimum cost. Priority would always be given to the social and economic sector for reason

of prudence as well as a Political necessity(consider the budget 2007-8 of India where an amount greater than the defence budget is proposed to be spent on waiving the bad loans of farmers). A finance minister while budgeting has to balance the total government revenue with that of tot al government expenditure plus the deficit financing (incase there is a shortfall in expected revenue).The non productive defence budget gets allocated by cutting other budgets of the economic / social sector or simply by additional revenue(tax). Therefore considering the developmental requirement the defence budget, almost, is pre-decided. Then there is the magic figure of 3% of the GDP that is touted to limit the budget. This is debatable. National security has no price and everything is subordinate to the exercise of national sovereignty. However this does not mean indiscriminate spending on defence. Here the cost is to be appreciated as an opportunity cost. Can the same objective be achieved by spending this amount on something else or dif ferent technology in a more efficient manner. At the national level this encompasses not just the defence expenditure but also politics & foreign policy where we may be able to cut (mutually) cost of defence by other alliances and agreements. However the minimum creditable deterrence will have to be maintained and therefore the figure of 3% has little relevance.

31. While arriving at the precise quantum of funds Richard Stubbings wrote about two problems in defence budgeting.

(a) Uncertainty about the threat, size and quality of forces needed to protect against the threat, and finally the cost of acquiring this force.

(b) Advocacy in rational defense planning – we tend to interpret events & elicit opinions which are in line with our personal interest and those of the Organization to which we feel loyalty . This is of interest in the field of defence because citizens lack expertise and access to classified data to form an independent judgment on military matters. The power of decision is centered on a few individuals in the defence establishment. The subjective opinions of these personnel carry an extraordinary weight. In addition there are other vested groups and political pressures that prevail on defence budgeting. While popular theory may contend that defence program and budget derive in an orderly 'top-down' manner leading from defined national needs and specific foreign policy goals, in reality the process is of ten exactly the opposite. Most often a bottoms-up approach prevails, where program & priorities are predetermined by competing bureaucracies who are adept in their ability to maintain status-quo. Thus the linkage between the actual defense budget and the national defense strategy can be mysterious – at times it may not seem to exist at all. In fact the entrenched lobbies are so strong that it is doubted if major changes can be initiated (those who thrive in the existing system will be the strongest opponent to change and in defence spending they have ample alibis to do so).

32. The budget consists of two parts the revenue and he capital expenditure. While revenue expenditure is understood and can be more precise in that it is required to maintain the establishment, manpower and the equipment the capital budget is different. For the purpose of this study we will focus only on the capital expenditure. Capital expenditure is for procurement of new equipment or replacement of existing one. The outlay is large and is normally carried out over several years and has to have concurrence of the finance ministry Therefore this must be planned where partial allocations are catered for and made annually It also has to be spread over various areas concurrently since tactical and strategic requirements both need simultaneous attention. This is as per the plans drawn up earlier. Budget is a tool of control as is well understood in study and practice of the science of management. If this be so, how do we effectively and efficiently use the budget to get value for each rupee spent. Obviously this can only be done by good planning, an efficient organization and unambiguous work procedures implemented.

33. **Responsibility for timely implementation -** The responsibility for exercising the budget needs to be institutionalized. Once the objectives have been set and their funding planned the objectives must be met. In case this is not done it implies that the earlier estimates were grossly wrong or that the procurement procedures are inefficient. In both cases the argument at a national level are unacceptable. The surrender of the budget year after year cannot be justified. Does it mean that our method of work is hindering the operational preparedness of national defence? An answer that the most naïve will not swallow. The natural inference is that there are vested interest that carry out a passive sabotage. In such cases pinpointing of responsibility would be next to impossible, however instituting transparency, adopting good procedures and fixed time limits for action at each stage would considerably reduce this. The responsibility should start 'top down'. The ministry of Finance should allocate the budget (capital) specific to each program. In case of any inability the valid and extenuating reasons are explained to the parliament and hence gain a collective responsibility of the leadership at the national level. The ministry of defence has the dual responsibility of working claims to the budget in consultation with the service headquarters or the Integrated defence staff and should not be a 'post office' between the ministry and services. The second role is to monitor expenditure and assume responsibility for surrendering funds. It must in its annual report and to the parliament present reasons for these surrenders. Secrecy on the ground of security in this regard has no credibility in today's wired world. The services themselves must reorient to ensure that the objectives are met once funds have been positioned. Not being able to then, meet the

objectives must be viewed seriously on the performance capability of the concerned authority.

34. **Monitoring & control -** This is the last and most important link in the process of defence budgeting. The process of monitoring must be implemented as a constant process. Timely and periodic intimation of the progress of the project ensure that mid-term corrective action is taken and if required policy is amended. Control as Stubbings says, is of two types the external and the internal control. The external control is exercised on the defence budget by the Finance ministry, the parliament etc. The internal control is achieved by formulating good policy/ procedures, operational control, sub-allocation of budget in order to delegate responsibility further and starting specific projects that have a comprehensive mandate. This ensures that lapsing of the budget do not occur again and again. It also ensures that there is a firm grip on the entire process and objectives are achieved on time. The ultimate aim of the defence budget is not the cost so much as that the right thing is made available in the right time.

35. **Practical implementation -** The practical implementation would include appointing a project director having under him product managers responsible for individual task/ systems. A review of the progress can be made to the project director who in turn will carry out a quarterly review of the progress with the DPB. Simultaneously and independently the seller would have a review meeting with the DPB so that the views are obtained from both sides in order to resolve kinks and bottlenecks experienced by the seller and buyer teams. For example in purchase of Tanks the project director can be a senior officer from armoured corps with project managers handling systems like Gunnery, automotive system, optics, communications and weapon systems. Additional managers will have to be appointed for quality control, warehousing & spares support and finally maintenance & Overhaul. At the contract formulation stage, views and functional requirements of all project managers will be taken and incorporated. Further purchase process starts from the suppliers premises when equipment is under manufacture. At this stage the DGQA have to associate to get in-depth knowledge. Also prior to dispatch the warehousing specialists have to be associated at the point of dispatch so that all spares are packed and dispatched in a manner that is in line with our distribution system and accounting to preclude chaos and losses later Lastly the policy on disposal of surplus, scrap and spares have to be spelt out to efficiently and economically dispose of the unwanted material worth crores of rupees that is accumulated year after year and occupies valuable space and incur expenditure on its accounting.

36. **Analysis -** The process of defence budgeting has been always under a veil of secrecy citing reasons of security. Therefore precise estimation and analysis is difficult except from

the figures available in the open forum. While the budget should be program based it is not so. In a program based budget the pre requisite is to have a national policy on security , national objectives, a long term defence policy to meet these objectives and a Budget commitment to back up this plan. The reverse holds true in the Indian environment. The budget is allocation of funds which may or may not have any linkages to the force capbility. While the ministry of defence projects the requirement to the ministry of finance it has no say in the final allocations which are passed on to the three services.

37. The defence budget is the largest recipient of the budgetary resources at 14 %. As per estimates of SIPRI it is the second largest importer of milit ary hardware after China spending almost US $ 30 billion. The paradox is that its indigenous defence industry contributes only 30 % of defence hardware when it has the technical capability, resources and requisite manpower available. It is puzzling that if this is the quantum of expenditure why efforts were not initiated to develop the indigenous cap ability (like food production). The import lobby is therefore more powerful than it meets the eye. The magnitude of the defence spending can be assessed as per the defence estimates of the last four years is given below.

Revenue and Capital expenditure and percentages thereof						
Financial Year	**Revenue Expenditu-re (%)**	**Capital Expenditu-re (%)**	**Total Allocation**	**% share Army**	**%age share Navy**	**%age share of Air force**
2005-06	48211.11 (59.9)	32337.87 (40.1)	51444.98	49.4	17.3	26.9
2006-07	51668.84 (60.4)	33825.8 (39.6)	854946.64	48.1	19.1	28.9
2007 - 08(RE)	54795 (59.2)	37705 (40.8)	92500	50.8	17.4	27.1
2008 - 09(BE)	57593 (54.5)	48007 (45.5)	105600	48.3	18.5	29.9

* RE = revised estimate BE = budget estimate. Source – Defence services estimates of concerned years. (figures may nor tally due to rounding off)

38. The method followed by the ministry of Finance is to add a percent age over the previous year's allocation to cater for inflation and if conditions warrant add a little more

and forward to the ministry of Defence. It will be seen that the percentages of allocations over the past have not undergone drastic changes year on year However the share of the Air force has increased due to high cost equipment followed by the NavyThe share of the army has come down. The positive aspect is that there has been a steady increase in the defence budget over the past years. Increase or decrease in the allocations by the ministry of finance has to be program based and intimated to the parliament.

39. The ministry of defence sub divides the allocations received to the services without altering the overall capital and revenue percentage as allotted by the finance ministry The changes made by the finance ministry to the projections received have no relationship to any proposed or ongoing project and it is left to the service HQ to determine this. To illustrate, details of projections and actual allocations for the past four years are as follows:-

	Revenue	Revenue	Capital	Capital
Financial year	**Projections**	**Allocations**	**Projections**	**Allocations**
2004 - 05	31915.01	27140.58	17202.25	8831.86
2005 - 06	30746.15	30535.46	13944.58	9421.14
2006 - 07	32617.51	31527.23	11941.28	10387.56
2007 - 08	35959.90	33269.16	13518.30	11621.02

40. The modernization plan of the defence services is dependent upon the fiscal commitments made in its support. Modernization is a long drawn process and may take a weapon system 10 years to be fully inducted, integrated and exploitedThe 15 year planning of the three services is an in-house exercise and without financial backing remains indicative and uncertain. Therefore budgeting remains an annual affair subject to revisions and re-appropriations with consequent lament that the allotted budget is not spentAfter 60 years of independence it is hard to believe that budget is not spent (in a planned manner and not by budget busting) if not by design? The fact that the 15 year planning is just a paper exercise is indicated by the results thereof.

India's five year defence plans (SP'S Military Year book 2006-07)

Plan	Plan period	Final status
7th Plan	1985- 1990	Finalised in Aug 1988. Actual budgetary allocation was Rs 58,188 Cr compared to the plan figure of 71,938 Cr. Representing a shortfall of 19%
8th Plan	1990-1995 converted to 1992-97	Not approved. Against plan of 14000 cr . the fin ministry allocated 107,000 crores ie a shortfall of 24%
9th Plan	1997-2002	Approved but with little linkages to the annual budgets
10th Plan	2002-2007	Still to be approved even in the fourth year of the 5 year plan period. Approved in the last year of the plan period.
11th Plan	2007-2012	Prepared but final fiscal allocations and expenditure to be seen Yet to be approved

41. The areas of concern that need attention are as follows:-

- There is no linkage between planning and budgeting. This creates uncertainty on what is the defence objective and lack of a yardstick to measure performance and defence readiness.
- There is a lack of awareness and hence involvement is restricted generally to lower levels. Debate at the higher level is on the levels of allocations and in broad terms only. Decisions are taken by the few and there is no debate or transparency for larger participation. This results in subjectivity and adhocism.
- The obsessive bogey of secrecy needs to be examined in the information age where secondary data, internet and study by analyst enable assessment of defence holdings and capability accurately. This is especially true when 70 percent of equipment is imported and open knowledge. Is the secrecy for home consumption?
- The objectives determine the required force level and the budgetary support to maintain the same, in this case it is not so. Force levels are decided in an incremental way and leads to obligatory expenditure which has to be catered for. There are sporadic exercises to review manpower levels which are related to existing workload rather than the military objectives and equipment required to be manned/ maintained.

- Defence equipment is inducted on life cycle concept. The cost of the equipment has cost in addition to the purchase price that includes the cost of maintenance, warehousing, manpower, overhaul and finally cost of disposal. This life time cost has to be evaluated before a purchase decision is taken. Merely going by the lowest tender is incorrect. This is an important input for the private industry for equipment sustainment.

- Three generations are required to be concurrently in service the 1st generation on its way for de-induction the 2nd being the main stay and 3rd being inducted into service. This can only be implemented if budgetary planning coincides with long term integrated perspective planning (LTIPP). If modernization/ updating is not carried out in a planned manner it leaves a gap and the easiest way to cover this is to extend the life of equipment. This is welcome only by the agency responsible to do so for obvious reasons. Really it is throwing good money after obsolete equipment besides adversely affecting the operational capability. Adhering to this policy further motivates private industry to develop next generation of the equipment. The Arun Singh committee has also made recommendations on these lines.

- Budgetary projections by the services do not coincide with allocations and is not necessary. What is important is the approach has to be same. The service HQ make projections based on capability requirements whereas the approach of the finance ministry is control of funds and allocations is based on last year expenditure plus ten percent. In the overall budgetary exercise the defence budget is almost pre-decided. The finance ministry must justify to the parliament areas against which cuts/enhancement of budget have been made and not follow a general thumb rule.

- Planning has been left to the services as a military matter and the ministry is only concerned about the resource allocations, whereas both aspects must go hand in hand. The priorities must be highlighted for instance funds must be allocated for eventual implementation of long term objectives like Network centric capability, Space capability, Capability of Cyber-war, PhD courses by officers, Automation in the defence services etc so that these programs proceed unhindered and industry can plan their production and cash-flows accordingly.

- Surrender of funds, or non utilization or re-appropriations is a result of ambiguity. This results in procurement of material in haste to beat the deadline or defer the purchase, both undesirable in any organization. Tentative efforts have been made to preclude surrender of funds by streamlining the procurement and evaluation procedures but result s are skeptical and would result in budget busting. It is a tunneled approach to a larger problem.

- Lack of long term policy is a major dis-incentive to promote the defence industry . The corporate sector cannot fathom the futuristic defence needs and hence no effort made to develop related industry. This results in dependence on imports.
- From the 6 th national five year plan the defence five year plan have become concurrent. A planning perspective over 15 year horizon was considered as per recommendations of the Arun Singh committee w.e.f. 1992-1997. These also get delayed approval and are not backed with budgetary assurances or commitments. The 8th and 10 th FYDP were never formally approved. Each acquisition case is questioned on it s need, quantity and rationale instead of having a broad policy approval of which this should be a part leading to delays. Gestation period of defence acquisition is large and the delay may not result in gain of strategic advant age besides, cause sub optimum utilization and exploitation in a fast developing world.
- There is an urgent need to empower HQ IDS, which is part of the MoD to plan the budget and to monitor it execution so that the nation get s value for each rupee spent. This will never happen unless a CDS is appointed.
- The present system does not share the long term vision with the environment thereby denying them the advant age of a planning lead time for product innovation. The user will certainly demand these when they become available to others and at that time it is not possible for R&D to catch up with lost time. Why the American defence do not complain about delays in new product s because the industry t akes the initiative to develop and present the armed forces a new product. In our case it is a situation of trying to catch-up and by the time we do it is obsolescent.
- The current method is not budgeting for a planned aim but a simple management of funds.

The above being a policy matter general observations only have been made.

Defence Acquisition

42. The terms procurement and acquisition are often used as one actually the difference is that acquisition refers to capit al purchases and procurement to revenue expenditure. Both are complex procedures for the purchase of materials and services. In government organizations it occupies a prominent stage since very large sums of money are involved. Accountability too is great as is the public gaze under which it functions. There are

extraneous pressures from interested groups of manufacturers, importers, Political groups, professional, Technical and financial views.

43. A simplistic view is to procure materials at the lowest cost. This may not always be practical as cost depends directly on quality. While cost can be quantified it is difficult to quantify/ measure quality. Further cost also depends upon the amount spent on after sales maintenance and cost of spares and training of manpower. The life of the item is another cost factor. Other factors that add or reduce the cost are payment schedules, credit terms, quantity discounts, level of technology etc. currently the procedure in case of capital acquisition is to have two quotations from select vendors allowed to bid. The first part is the technical bid wherein the bidders compete with one another to offer the best technological product and its performance features and the second is the commercial bid where the price is quoted along with qualifiers if any. Assuming all bidders meet the technical criteria the bidder with the lowest cost generally wins the contract. In theory this may be plausible but in practice other factors may weigh against it these are like:-

- Ability of the bidder to actually execute the order
- Execute it on time
- Have in-house manufacturing facility or trading capability only.
- Past performance and financial capability
- Post supply cost of training
- Maintenance cost
- Rejection rate / rework required

44. The second alternative is to take a holistic view of all factors and it is possible that the lowest bidder (L1) may not in fact be the lowest in the long run. Therefore costing should be done taking into account the life time cost of an item.

45. The third alternative is to arrive at a reasonable quality by paying a reasonable cost. This may be applicable in certain cases. The procedure works on the premise that all other conditions being equal the highest and the lowest bidders are rejected as being aberrations. Thereafter a simple average is calculated of all the prices quoted and this is assumed to be closest to the reasonable price. The bidder closest to this price is made the first offer.

46. Whatever the methodology employed the system needs a dedicated organization staffed with experienced personnel in the area of analyzing and procurement processing with contract servicing. Their training needs to be continuous to be at par with current financial and commercial practices. In the MoD staff is not selected by their experience in spite of having such people and this is detriment to the organization. The GoM in their report post Kargil have recommended a National defence university (NDU). Under the NDU the College of materials management could be chartered with the t ask of research on the application and practices in Purchases.

47. The purchase procedures have been formed both for acquisition as well as procurement vide the DPP 2008 and the DPM 2008 respectively For acquisition the defence acquisition council has been formed and at the low end of expenditure the defence procurement board formed. These institutions will have a continuity and institutional memory . The acquisition policy in defence is complex and is usually drawn out over a long period due to evaluation of the product, its testing under all weather conditions, technical and commercial evaluation and finally negotiations. The induction itself may take a long time.

Defence Procurement Procedure (DPP)

48. Defence Procurement Policy was formulated in det ail in 2006 and it was revised in 2008. The entire procedure is lengthy and available on the website of the Ministry and hence not dwelled upon here. The salient features are given below.

49. The DPP 2008 became effective with effect from 1 Sept 2008. The highlights are as follows:-

- Procurement will be time bound
- A defence procurement network will connect all concerned agencies to build up a data base and information system.
- Greater transparency and competitive. RPF will be available on the MoD website giving vendors adequate time for prep aration. Broad det ails of trials will also be given. Results of trials and reasons for disqualification will also be given.
- Multi vendor system being encouraged and discourage single vendor system.
- Provision of Of fset banking included. (Though 2 years is less, probably will get revised af ter experience). Not applicable in case Indian share in a comp any is greater than 50 %. Fast track acquisition exempted from offset.
- Offset banking credit should be transferable.

- Powers of tech evaluation committee enhanced.
- Integrity pact will be signed by the DPSU with their sub vendors.
- Inclusion of companies for development of TOT.
- Financial powers of military up to Rs. 50 crores and Defence Secretary up to Rs. 75 crores granted.

Short Comings

- Foreign Exchange rate variations not made applicable to private industry whereas applicable to DPSU.
- Variation in terms of payments between Indian & foreign vendors. also between Private industry vs. DPSU.
- Indian offset partner required to pay more tax on work done against offset, this could be exempted
- Pre qualification of firm in terms of financial & tech capability should be a criteria for issue of RFP/ RFI.
- UPGRADE programs should BE RESERVED FOR INDIAN FIRMS ONLY.
- DOFA kept out of offset banking , so who will study the efect of offset banking and prepare proposals to improve them
- DOFA has no powers and must have teeth to make offset a success.

Recommendations

50. The recommendations are as follows:-

- Policy on national security policy has to be articulated in the overall interest of the country. Planning should be carried out between the Planning commission, Finance ministry and the Defence ministry, Followed by the defence policy.
- The LTIPP of the services must be integrated in the planning of the finance ministry and Planning commission.
- Capital acquisition must be known from the stage that approval is granted for it. (Most of the secret information is available worldwide in trade journals and research papers, and is classified as secret for home consumption).

- The defence minister must have the final say on planning and executing the defence budget.
- The budgeting and procurement procedure is based on the works procedure and needs to be reviewed.
- The CAG must carry out an *annual review* of the defence spending and report the same to parliament.
- All defence acquisition of capital equipment must be taken up as a project under an appointed leader and a supporting team who will be accountable from induction to total deployment. Further allocation of funds or withholding the same must be conditional on performance.
- The procurement system based on L1 vendor needs to be revised.
- Defence procurement must provide an even playing ground for all suppliers less foreign bidders.
- The recommendations of the Group of ministers and the Arun Singh Committee, Ramrao committee and Kelkar committees must be implemented.(relevant abstracts are at Annexure 1 and 2 to this chapter.)
- As an interim measure the five year defence plans must be approved along with the national Budget of the first year of the plan and revised each year thereafter.
- The catalyst to achieve the above can only be a CDS who must be appointed without further delay.
- Effect of variations in rate of exchange of foreign currency must be spelt out in the DPP.
- DOFA must be an institutional body with a separate secretariat and budget.
- Defence forces must be equipped on capability based performance.
- Industrial coordinating agency at the apex must deal with issues of offset banking.

Annexure 1

Relevant Abstracts of the Expert Committees on various components of DIB

Kelkar Committee Report

Ministry of Defence appointed a committee to examine and recommend modalities for the integration of users, Defence Ministry and the Indian industry in the acquisition of specific product required by the armed forces.The committee headed by DrVijay Kelkar started its deliberations in July 2004 Part 1 of the report was submitted in April 2005.The key recommendations of the committee concerning DIB were as follows :-

1. Prepare 15 year plan forming the basis for acquisition programme.
2. Identify entry points for the private sector in the acquisition programme.
3. Accredit Raksha Udyog Ratna (RUR).
4. Policy frame work to promote participation of SME s in defence production
5. Enhance defence R&D.
6. Promote transparency in decision making.
7. Optimum utilization of capacity.
8. Armed forces to make public version of Technology perspective road map.
9. Amend the constitution of Defence Production Board to have reps from the industry.
10. Inclusion of rep of CSIR and ISRO in the Defence R&D Board.
11. Ensure availability of funds for R&D for the private sector.
12. In medium and low technology areas ,DRDO should off-load as much as possible to industry.
13. MoD should adopt a DARPA like model for giving R&D for higher technologies to the private sector and other scientific institutions.
14. Up grades of existing platforms which have adequate lead time should be done preferably by indigenous industries.
15. Introduce the concept of “shared development cost” on the lines of ISRO model.
16. Create a National Centre for strategic information Technology as an autonomous , no profit management oriented professional body to conceptualize IT related Defence projects.

17. Creation of a "Strategic Defence Industry Fund" for non lapsable pool of resources exclusively for "Make" projects.

The recommendations are very thoughful but only few of the recommendations have been implemented so far.

Annexure 2

Rama Rao Committee Report

Extracts of the report related to DIB are as follows:-

1. Emerging technologies that need to be focused in are Biotech and aerospace. Design seems to be ready to replicate what has been achieved by the IT sector in India.
2. The armed forces depend upon 70% of bought-out arsenal.
   In basic science India has declined to a rank below 22 in the world, Below China, Poland, S Korea.
3. PhD in engineering India-700, China- 7000, Taiwan- 4000, S Korea-1400.
4. Scientist in R&D per million of population are: India-120, China-633, S.Korea-2979, japan-5085.
5. Expenditure on R&D as a percentage of the GDP. India-0.8%, China-1.2 %, USA- 2-2%
6. In India availability of talented personnel is not a problem but their management and reward is.
7. There is a requirement of large scale Public and Private partnership.
8. There is an urgent need for an active S&T policy without which India will decline to the status of a technological client loosing independence by being dependent on other countries for technology rather than be an innovator. It will become a source of blue collar job provider and of services. As a global player India will have to set the rules of global engagement advantageous to its interests.
9. Critical sectors of science & technology are

- Nano tech
- Material processing-fabrication, including electron & photon scales.
- Creating and fabrication of bio materials for processing of living matter on the scale of micro-organisms and lower (down to DNA).
- Technology for the extraction of energy supplies and raw materials from sources including sea bed

- Computers and communication technology.

* Incidentally – China has a policy wherein all ICT equipment has to be inspected, tested, and certified secure from national security angle. Further there is a complete prohibition of remote electronic-servicing of communication. Services that require server, switches, network management communication equipment are required to be security certified.

10. Armament technology is required in the following areas
    - Better sensors
    - Faster computers
    - Stand-off weapons
    - Enhanced transparency digitalization of the battle field.
    - Seamless communications
11. Frontier technologies required are:-
    - Miniaturized / net worked, multi physics, hyper spectral sensors.
    - Robotics
    - Long range precision weapons
    - Info/ net warfare
    - Mini/ micro satellites, missiles and UAV's
    - Binary bio-weapons
    - Smart mines
12. India needs to adopt an approach of proactive persuasion towards Pak and a policy of competitive capability development both economically and militarily. Adopt a policy of jointmanship.
13. There is a need to develop cap ability of containing terrorism and proxy war who use fourth generation weapons and modern means which are adding to the complexities. (*Disaster management would therefore have to be organized into a more effective entity to function besides the security forces in the overall plan of fighting terrorism and proxy wars).*

14. Future wars will require the following in areas of communication:-

- Interoperability and a 'system of systems'
- Cross Service connection between sensor and shooter (weapon)
- Command control and communication will need greater inter-service bonding.
- Nuclear and international relations are pushing conventional conflict s in the background and therefore rise of sub-conventional & asymmetric threat has increased.
- Develop a system to attack the enemy's information system and safeguard our own.
- Asymmetric warfare means greater understanding of the strategic appreciation and operational planning.
- Thwart nuclear blackmail by non-state actors as a strategy.
- New challenges make it dificult to identify the adversaries thus making traditional threat based planning obsolete.

15. Hardware imitative – India should get/ build on chip technology & design to include state of the art (0.25 micron) silicon fabrication. Should aim to build world class fabrication that can use 90 nm technology with 200 mm wafer and the approach should be studied against state of art 300 mm fabrication.

16. Technology management issues:-

- Defence tech commission be set to oversee make/ buy having cross representation or a Defence Technology Board under NSA, CDS or a separate scientific advisor to the Defence minister.
- Each service should have its own technological advancing mechanism in the form of a 'technical and system command'.
- Synergy between defence and civil in areas of aeronautics, Naval technology, and establish the structure for aero and naval tech.
- The concept of offset clause is welcome.
- Separate structure could be established for LCA, Agni, Arjun, Sagarika under a separate head. Give LCA project to the Air-Force and HAL.

- Exports of defence products be encouraged.
- Develop separate technical centres for developing technology like nuclear and Space.

Fourteenth Report of Standing Committee on Defence "2006-07".

This report was presented to the Parliament on16.3.2007.The focus of the report is on DRDO. Most of the recommendations made are yet to be implemented.

Chapter 6

Sectoral Study of Selected Products / Technologies

Rationale for Choice of Sectors for the Study

1. The choice of sector / technology for the purpose of any study can cover a very wide canvas. The desire to have everything vis a vis the need to balance the focus of attention to the attainable has to be maintained. Care has to be taken that the selected technology has a future perspective and has the potential of enabling a quantum leap forward to match the best without having to through the process of re-inventing the wheel.Therefore to select the crucial technology which will imp art the minimum cap ability required is a deliberate exercise and during the course of the study the following areas for application were selected.

- **Land systems-Infantry, Armoured vehicles & Artillery**
- **Marine systems – Warship building**
- **Aerospace systems –Missiles, Satellites& aircrafts**
- **Electronic & Cyber war**

Land Systems

Combat Vehicles

2. In land based operations the most critical p art of the milit ary organization is the mechanized force. This arm of the army is responsible for forward reconnaissance, flanking, defensive operations, and neutralizing critical strongholds in the depth thereby breaking the will of the enemy to fight. While the mechanized forces achieve this they are not able to physically hold ground for which the Infantry follows them to do so. Logically it follows, that the natural ability of a human in an Infantry soldier be enhanced and supplemented to enable him to ef fectively communicate with the environment, and fight the enemy with greater protection to self to defeat him.

3. The genesis of mechanized or tank warfare started with attempts to break the deadlock of the trench warfare where long artillery battles followed by infantry operations yielded or lost a few hundred yards of ground. In order to restore mobility to the battle field (earlier enjoyed by cavalry), bulldozer with armour protection was used to roll over the trenches. Since then it has evolved to reach its present state. The development of the modern Tank has been a history of achieving opposing technology i.e. armour protection vis a vis means to defeat it. As there are advances in technology so are the methods of defeating armour and methods of protection.Advances in composite materials and nanotechnology are likely to witness light weight armour thereby enhancing the mobility of Tanks. Powerful Sensors, Radars and counter measures will also impart capability to detect incoming missiles and measures to counter or deflect them Anti-Tank measures are likely to have smart munitions and missiles with low Radar signature and stealth technology with high lethality and GPS assisted precision guidance. There are likely to be advances made in unmanned armoured vehicles and anti-tank weapons remotely controlled. There would be a need to remove night blindness to preclude this distinct disadvantage in armoured warfare. Simultaneously the threat perception will also dictate the use of technology. Thus we witness in an environment of asymmetric warfare the requirement of armoured vehicle with high mobility in an urban terrain, necessitating use of wheeled armoured vehicle replacing the traditional tracked vehicle. Therefore the need is to forecast the future technology in armoured warfare and develop the related basic technology required.

4. The following are the areas of modern or emerging technology related to armoured fighting vehicles, mastery of which will impart superiority to the country.

- Use of reactive armour or other innovative material for lighter and better armour protection. *(Composite material & metal research)*
- Laser guided ammunition and other standoff weapons. *(laser & guidance (GPS) for weapons)*
- Night sights using thermal imaging up to a range of 3000 meters *(Image intensifiers)*
- Laser infrared illuminator *(Image illuminators)*
- Target detection and tracking radar (*Radar technology)*
- Hermetic sealing and ventilation of tanks for operation in NBC warfare *(NBC capable vehicles)*
- Engine technology producing up to 1000 HP of power and development of hybrid engines and electro-drive technology. *(Engine technology)*.

- Innovative exhaust and ventilation system to reduce thermal signature *(Emission and acoustics control)*
- Day and image intensified night sight *(Image intensifier)*
- On board Enhanced Systems-control. *(Control systems)*
- Robotic engineering for unmanned combat vehicle. *(Robotics)* .... (Having 6 wheel and self inflatable tyres. It is a hybrid powered vehicle. can be tele-operated and can withstand occasional moderate crash. It can surmount rough terrain to a large extent.
- GPS navigation and communication system *(Secure global positioning systems).*
- Reduction in onboard modules to reduce weight and enhance maneuverability *(Electronic miniaturisation).*

Technology for Infantry

5. As the technologically driven warfare evolves it necessitates a change in the methods of waging war. Thus we witness the emergence of RMA and 4GW exploiting technology. Emerging trends indicate that armed forces would have minimum cold st art time, would need to have highest mobility and use of precision standoff weapon systems. Operations would be carried out in a transparent manner in full view of the world, ruling out any collateral damage. On the reverse side it also means minimum casualties of own troops Bad planning, lack of necessary ware with all or poor intelligence will be in full public view and shortcomings if any will be ruthlessly accounted for and questioned by the Public from the Political and Military leaders. These are the paradigm shift that will have to be catered for in the emerging battle environment. Therefore as a corollary it follows that own troops must have minimum casualties and will have to be equipped using st ate of art technology in protection, information awareness, effective personal weapon and immediate evacuation of casualties.

6. The Infantry soldier is expected to be the most versatile element of any armed force. The outcome of and armed operation is dependent upon his ingenuity to exploit the weapon or other resources available at the opportune moment. The entire defence superstructure is based on this assumption. In the field the soldier is vulnerable first *psychologically* and then *physically*. In a NCW environment and asymmetric warfare the prime need is to integrate the soldier with the highest commander with ef fective communication so that complete situation awareness is there this caters to the first vulnerability. The second is to provide personal protection and offensive capability. A human is only capable of carrying so much load and to maintain peak efficiency this must be kept to a minimum. Further he must be

able to see better during night and day.

7. Developed countries have initiated the process of equipping the infantry soldier using the best that technology has to offer and this will steadily be enhanced. The French army has taken the lead in introducing the future soldier system called FELIN (Fantassin a Equipments et Liaisons Integres). The US army has the Land warrior, the Germans have IdZ (Infanterist der Zukunft) and the British army the FIST (Future Integrated Soldier Technology). The Indian army proposes to introduce the F-INSAS. Future wars will not be in the conventional form and will be local conflicts of short duration. Greater number of conflict will be fought by non state actors using technology and under asymmetrical conditions. This again underlines the need to have a military force that is built on capability rather than based on threats. The need for any army would be to equip a minimum level of troops that would meet its current requirement.

8. In the Indian context considering our environment as well as past experience it would be more practical to organise our fighting formations as a fully self contained unit. The soldier will be equipped as per the role he is to play i.e. Leader of a unit, Sub-leader, Rifleman or Sniper. The mission will also dictate the technology used say for NBC environment, fighting in built up areas, jungle warfare, operations in high altitude etc. this could be restricted to select formations only. Technology required to equip the infantry soldier to meet all foreseen condition are as follows:-

- Personal light weight armour protection *(material research & nano technology)*
- Radio communication on multiple frequencies and Nets *(miniaturisation of radio and video technology)*
- Enhanced power for communicating *(high quality batteries or alternatives with re-chargers)*
- Portable night and day image intensifiers. *(image intensifiers)*
- Light weight weapons and ammunition *(small arms technology with inbuilt laser range finder)*
- Light ammunition *(ammunition technology)*
- Navigation and target designation *(portable GPS and target designator for bringing down heavy fire)*
- Endurance enhancing *(research in biotechnology to enhance resistance to disease, reduce pain and increase stamina)*

- Clothing that protect him under a variety of conditions. Thermal, Fire and water resistance fabric *(research in textiles).*
- Footwear that improves protection from mines and shrapnel (*composite materials development).*
- System to identify friend or foe instantaneously *(Sensor technology)*

9. The army has initiated the process of equipping the soldier on these lines to be able to meet the challenges in the emerging multi spectral environment. The project F-INSAS stands for 'Future infantry soldier system'. Needless to say that this will be implemented incrementally while simultaneously upgrading the education and capability of the soldier to exploit the system. It would also require the infrastructure to implement and support the system. Details of the most advanced system FELIN has been deployed by the French army. Systems developed or under development by other countries are more or less on similar lines.

Artillery

10. The next important portion of the land systems is Artillery. Artillery consists of the Ordnance and the other allied systems which complete the Gun. It also has a power drive to tow the guns. The areas of focus for technology development is the ordnance, which is similar to that of aTank hence not required to be duplicated. Further the focus of artillery is shifting to the use of missiles which is being considered hence not covered under this head.

Marine Systems

11. The Naval force consist s of warship s of dif ferent types depending upon their role supported by a complement of Sub surface vessels and technology . It is further complemented with an air arm consisting of aircraft and helicopters to enhance the reach and capability of the Ships. The importance of the Navy is best described in the words of the American president Theodore Roosevelt '*A good Navy is not a provocation to war It is the surest guaranty of peace*'. This cannot be truer for the current Indian situation. Having a coast line of 7500 Kms, it occupies a strategic position in the Indian Ocean across which majority of the world trade and oil shipping is carried out. Emerging as a major regional economic power India has to safeguard its interests by securing its boundaries, shipping lanes and using the Navy for force projection as deterrence. In the current environment the Navy, besides a military force also has a diplomatic role as part of the national foreign policy.

12. The capability of the Indian Navy should be safe guard sovereign assets in the Arabian Sea, Bay of Bengal and the Indian Ocean. It should be able to prevent poaching in its territorial waters and safeguard its international shipping and trade. This needs a blue water capability and a reach extending from the Persian Gulf to the Malacca straits. The Navy therefore forms a critical part of the study to assess its need for emerging technology

13. India is not a major player in shipbuilding industry of the world. The composition of shipbuilding in the world in 2005 was as follows. It relates to both defence and civilian sectors. Most of the manufacturing facilities can be interchangeably used in this industry for defence and other purposes.

COUNTRY	MILLION G.T.	% SHARE ON DWT BASIS
South Korea	20	35
Japan	17	33
China	7	20
Europe	5	8
Others	3	4
Total	52	100

Source : Internet

14. Production of India appears under "others", share of which merely stands at 4 percent collectively for many countries. While military shipbuilding is still considered at its infancy stage in Asian shipyards, they are growing at a rapid rate. Korean shipbuilders are major suppliers of patrol and supply boats for navies. The overall position of important military navies are as follows:-

- The Republic of Korea Navy has embarked on a vigorous shipbuilding programme to become a substantial blue – water navy by 2020.
- The US Navy as of February 2007 has around 276 ships in its fleet and its shipbuilding plan calls for 326 ships by 2020. They plan to add a whole new class of multi – mission modular vessels called littoral combat ships.
- Russian Navy is expected to become the world's second largest in 20 years time. They have enormous capacity and technology for military shipbuilding. They are main collaborators and supplier for the Indian Navy.

- Indian Navy operates more than 155 vessels including one Aircraft carrier. They have a programmme to add new vessels to their fleet.
- China has experienced rapid growth in it s shipping industry. In 2005, it had third place among the shipbuilding countries of the world. However, its aim is to become second by 2010, and the leading country by 2020.

15. Geographical spread of shipbuilding in the world by revenue is as follows (in % terms) :-

Country	% Of Revenue
North America	58.4
Europe	26.4
North Asia	12.8
South America	1.6
Oceania	0.3
Africa And Middle East	0.2
South East Asia	0.2
India And Central Asia	0.1

16. Segmentation of products and services in percentage is as follows: Submarines 49.1 %, Surface combatants (including boats) 22%, Auxiliary ships and Boats 16.3%, and Repair and other services 12.6. % It may be noticed that major share of market belongs to submarines, for which India has very little activity.

17. India has five major shipyards operating under the control of the Government. These have a predominantly civilian manpower and are Unionized, therefore to run them as a military organization may not meet the ends of exploiting installed capacity. There exists a strong case to convert these Shipyards into corporations to facilitate their management on corporate lines. As a government organisation they are handicapped in providing adequate incentives, marketing, innovation and attracting the right t alent to compete with leading shipyards of the world. These shipyards are:-

- Cochin shipyard ltd; Cochin
- Mazagaon Docks ltd; Mumbai

- Hindustan shipyard ltd; Vishakhapatnam.
- Garden Reach Shipbuilders and Engineers, Kolkata
- Goa shipyard Ltd; Goa.

18. They are involved in shipbuilding and other activities for both the defence and civilian sectors. The industry is geographically dispersed along the Indian coastline. Apart from production, these companies are involved in ship repairing activity.

19. The product range of these companies in the area of shipping is : Advanced off – Shore Petrol Vessels, Missile Craft, Fast attack craft, Off – shore Supply – cum – Standby Vessel, Sail training ship, 300 Passenger Vessel, 650 Ton Oil tanker, landing craft, Survey Vessel, Frigate, Missile Corvette, Fleet Replenishment Tanker, War ships, Submarines, Merchant ships upto 30,000 DWT, Bulk Carriers, High Bollard Pull Tugs, Air defence ship, Deck machinery, Marine Engines etc. Other products include: Dredgers, off – shore Plat forms etc. The largest ship that can be limit is 1,10,000 DWT at Cochin shipyard. It can repair ships up to 1, 25,000 DWT.

20. The current shipbuilding capacity in India is only 2,81,000 DWT , which is quite undersized according to global standards. According to KPMG, it is expected to go up to 4 million tons by 2012. Pipavav shipyard, the upcoming yard in Gujarat is rapidly becoming one of the largest shipbuilding complexes in the country It started production in February, 2007.

21. Main forte of Indian shipyards lies in medium and small vessel segment but planned investments are targeted towards construction of larger and more complex vessels. L& T is planning to build very large crude carriers. Pipavav shipyard has orders for 12 panamax bulk carriers for a Greek company. Indian shipbuilders have an order book of about 240 vessels, worth Rs. 18,000 crore ; 70 % of these orders are for foreign flag vessels.

22. There does not appear to be a substantial R & D activity in these companies. In fact, they are not expected to involve themselves in this activity . Concept design and sketch designs are provided to them.They do detailed designs and production drawings. R & D is the responsibility of the Directorate of Design and Development under the Ministry of Defence, and the Directorate of Research and Development (DRDO) under the same ministry. Modern means of production are being used in these shipyards, but not as extensively as in the Chinese Shipyards, since volumes involved in India are low . Considering the size of this industry, China has over 120 shipyards as against around 32 in India. Capacity of the Chinese shipyards are much larger as compared to the Indian shipyards.

23. Participation of private sector in the shipbuilding industry in India is insignificant at present. This industry is labour intensive and suit s our country. Considering the current demand in the Indian and the world market the private shipyards like:ABG, Bharti, Tebma etc are trying to expand their facilities. New players like: L& T, Goodearth, Mercator, Adani group and others have been trying to set up new shipyards. The government has also announced setting up of two new shipyards.

Modernisation & Development of Naval Technology

24. Modernisation of ship building is a complex effort that combines multiple technologies into a coordinated and interrelated system. Therefore as technology develops in any area it finds application in upgrading the function of a battleship, hence frequent requirement of dry docking for maintenance and refitting. Since naval ship s are required to undert ake military operations inherent requirement of stealth, speed, reliable communications, management of magnetic & underwater signatures assume greater impodnce. The major components that contribute to complete the ship platform as a war machine are:-

- War Ship design and hull form.
- Communications
- Propulsion system
- Radar & tracking systems
- Miscellaneous

Warship Designing

25. War ship design is an advanced technology and is being carried out by Government agencies in India. Indigenous ship construction activities have basically utilized materials like low alloy steels, carbon steels, cast iron, aluminium, tit anium and copper alloys. Emerging technology trends in warship design, material sciences and stealth technology are set to revolutionize warship ship building, providing platforms with better speeds and sea-keeping qualities, higher equipment package density, enhanced stealth features and reduced maintenance features. The important technologies that find application in shipbuilding are as follows:-

- **Shipbuilding Steel**: For surface ship s, steel has been developed recently . However, for sub-marine construction it is still being imported.
- **Composites**: Composites have a variety of uses in shipbuilding. They need to be

developed for items such as door , hanger shutters, masts, super structures etc. They are needed to provide stealth and reduce top weight of warships.

- **Cladded Metals**: Cladded steels are excellent materials with both strength and chemical resistant properties. These are particularly suitable for battery storage compartments.

- **Coating Materials**: The development of silicon – based coatings with natural antifouling against will be required.These coatings are important for life extension, reducing drag, increasing speed and extending range.

- **Materials**: A variety of materials are required for ship construction. These range from structural steels to composites and include insulation materials, deck covering materials, materials for piping and fixtures, main machinerysonar domes and paints for surface protection. Until recently , India was importing steels for warship construction. While this situation has been remedied to a large extent, continual R & D effort is required for developing emerging exotic materials, composites and panties. Position regarding some of the items is explained below:-

- **Nano – Phase Materials**: Nano Materials is an area of prime importance for future naval applications. Magnetic nano-materials may offer improved performance for magnetic storage applications. Nano material could be combined with nano – scale electronics to produce a new class of sensors able to achieve ultra – high speed and low-power dissipation. Some of the newer materials like: monoplane materials, smart materials, heterogeneous materials, super conducting materials, high temperature materials and functional materials have high potential for naval applications.

- **Miscelaneous** - Development work is needed on a number of other materials like: polymers, titanium, metallic foams, anti-fouling materials, stronger easily weldable less expensive steels for ship and submarine hulls, and special materials required for super conductors etc.

- Materials to be used for future naval engines should have reduced weight, increased temperature capabilities, improved mechanical properties and better corrosion and oxidation resistance. Such high performance materials include organic matrix composites, titanium alloys, and inter-metallic compounds.

- Improved magnetic materials will be required for magneto-optic devices and high sensitivity, low cost magnetic sensors to be used as magnetometers, radio-frequency antennas, and biological and chemical sensors.

Communication Systems

26. Communications is the lifeline of naval systems operating over vast distances over sea and underwater in all weather conditions to conduct operational and strategic missions. Secure communications are required between naval forces, land forces and the Air forces. These operate through satellites and global positioning systems in order to enable quick decision making and reduce time between the sensor and shooter

Propulsion

27. The Indian Navy currently employs three conventional propulsion modes i.e. steam plants, diesel engines, and gas turbines. Indigenously manufactured steam turbines of BHEL and main propulsion diesels of Kirloskar Oil Engines Ltd. are already in use onboard ships. The current naval power propulsion techniques lead to stealth – related issues (reduction in acoustics and IR signatures), indigenous development of gas turbines and non – conventional modes of propulsion. The fuel cell is the emerging technology as a non-conventional energy source for main propulsion and prime movers.

- **Gas Turbines**: Presently, all gas turbines fitted in Naval ships are of foreign origin. Indigenous initiatives taken in this regard include the following:-

  (i) General Electric's gas turbine is being inducted on the basis of its licensed manufacture in India with progressive. Increases in indigenization.

  (ii) Development of a fully indigenous Kaveri Marine gas turbine by GTRE, Bangalore.

28. There is a need to develop indigenous gas turbines in the range of 11-15 MW and 20-25 MW for fitment on board future ships as main propulsion units. Adequate emphasis has to be laid on development of gas turbines with enhanced aero-thermo-dynamics.

- **Diesel Engines**: Development in the field of diesel engines is driven by stringent environmental regulations and requirements of multi-fuel operation and long service life. More-over, development may be required for achieving better power to weight ratios and better torsional damping characteristics, across the entire power range of the engine.

- **Nuclear Power Propulsion:** Nuclear power presents the ultimate in "Air Independent Propulsion" (AIP) solution affording high speed, mobility, autonomy and submerged endurance limited only by stores capacity and crew fatigue.

Development of nuclear power propulsion plants may be required for the surface combatants of the navy.

Propulsion

29. Propulsion of all naval platforms are at the heart of their success. Developing technology manifest in their application in naval ships. Some emerging technology are :-

- **Fuel Cells**: Fuel cell technology is receiving world wide attention as it provides a viable AIP solution. The fuel cell power p acks may be developed for submarine main propulsion as well as energy sources for various prime movers. Many types of fuel cells are there like: Proton exchange membrane fuel cells, Direct Methanol fuel cells, Molten carbon fuel cells, PhosphoricAcid fuel cells, to mention a few Of all the AIP systems under development, the phosphoric acid fuel cell is widely accepted potentially as the most viable solution. It consist s of an anode and a cathode made of a finely dispersed platinum catalyst on carbon and a silicon carbide matrix that holds the phosphoric acid electrolyte. It has potential for providing high power outputs, suitable for naval propulsion systems including remote vehicles.
- **Electrical Propulsion**: Electrical Propulsion technology provides higher efficiency, improved survivability and lower noise signatures. Commercial shipping has already adopted this technology. It is now being adopted for warship application. Most of the elements required for warship applications are available in the international market and indigenous commercial market. The progressive development in fuel cells and super conduction technology will make the electrical propulsion option more attractive.
- **Low-Noise Gearboxes:** Noise generated from a gearbox contributes considerably to the overall noise level of the ship. Techniques such as finite element analysis should be developed to design comp act and silent gearboxes. Advanced manufacturing techniques, metallurgical processes and materials are required to be developed to meet silent gearbox standards.

Radar & Tracking Systems

30. Radars are an important constituent for tracking and engaging unfriendly targets and there is a constant effort to enhance their range and capability to track targets as early as possible.

- **Stealth Materials** : Development of stealth materials and paints like radar absorbent materials, Radar transparent materials, Radar opaque transparent materials need to be pursued further Development work is also needed on RCS reducing techniques like camouflage screens.

Other areas for R & D:

31. These are areas covered by the following type of technologies:

- Computational Fluid Dynamics for low noise fluid flow in ducts and pipes.
- Motion control systems / motion interceptors for roll and pitch stabilization for naval platforms.
- Air lubricated bearings for use in high speed turbines.
- Manufacturing technologies for producing / machining of parts used to build naval ships.
- New fire suppression technologies to replace Halons, as per Montreal Protocol.

Ship Automation

32. New construction ships are already being filled with "Integrated platform management systems" for control and monitoring of all plat form machinery and systems including propulsion, power generation and distribution, auxiliaries, damage control, steering and stabilization. Embedded processors will penetrate equipment to make them intelligent "appliance". It enable the equipment to indicate pre-failure warning and specify rep air instructions. Composite systems control technology is being widely used in India in the Petroleum refining industry and steel plants, this can be further developed indigenously for naval ships.

Sensors

33. The Indian Navy uses electromagnetic, acoustic, mechanical, environment al, and temporal sensor systems for surveillance, acquisition, and assimilation of information related to location, movement, and identification of friendly , neutral, and hostile forces. Sensors are also used for improving geo-spatial awareness, viz sea conditions, bathymetry as well as environment, weather and space characteristics. Acoustics remains the most dependable means of under water surveillance. Sonars will therefore be the main sensors for detection of below water bodies like submarines, mines etc. However, the propagation medium and its characteristics, such as temperature, salinity , wave p attern, bottom profile, ambient noise etc impose constraints which have to be overcome by utilizing novel transmission and signal processing techniques.

Communication

34. Indian Navy' s can truly become a blue – water navy in next 15-20 years if naval commanders at sea are able to synchronize and integrate high-tempo operations anywhere in the world. It would require global end – to – end information exchange among the units as a critical mission cap ability, and force-multiplier for worldwide readiness, mobility , responsiveness, and operations. This information exchange would need to be provided by a network of ef ficient communication systems, using satellite. Satellite communication should preferably be indigenous in the long run including the Global positioning systems (GPS).

Required Capabilities/ Changes

35. Essential capabilities / improvements required for the Indian Navy in the shipbuilding field are as follows:-

- **Ship Construction**:

  (i) Efficient hull forms for higher speeds and better sea-keeping capabilities.

  (ii) Adequate stealth features

  Various options have been mentioned under "Warship Design" earlier.

- **Materials**: This very import ant area. A number of new materials of construction have been developed for shipbuilding industry in the recent times.They have been briefly described earlier . They help in reducing weight, increasing temperature capabilities, improving mechanical properties, improving stealth qualities etc.They need to be introduced in the new ships.

- **Propulsion**:

  (i) Improved Gas Turbines and diesel propulsion systems for ships.

  (ii) Electric propulsion systems.

  (iii) Air Independent propulsion systems

  (iv) Acoustic noise reduction in machinery

  (v) Bulk energy storage device for short and long duration.

  (vi) Detection of chemical and biological attacks and protection, decontamination etc.

- **Gap in Technologies for which R & D is desired:**
    - Sophisticated sonars for anti – submarine defence
    - Long range and precision radar
    - Modern fire control systems for: Guns, Missiles, Torpedoes etc.
    - Material technologies for hull to make it lighter
    - Material technologies for "stealth" (paints etc)
    - Material technologies for electronics for Chips, Radars and Sensors
    - Opto electronics for sensors, thermal imaging etc.
    - Nanotechnology for possible applications.

The gaps have been explained in details under the heading "Warship Design" earlier.

Gap in Backward Integration for Manufacture of Ships:

36. Since import substitution is an ongoing process, latest position has to be ascertained from the shipyards. However, major items of import are: propulsion systems, navigation and sensor suites, and major weapon systems. Many materials of construction and parts are also imported. For some of the materials, quantities required are not large enough to justify economical manufacture in India, though technology exists in India to handle the same.

References:

- *Indian Navy science and Technology Road Map – 2004 Govt. of India, Ministry of Defence.*
- *Annual Report 2007-08, Ministry of Defence, Govt. of India.*
- *Public Enterprises Survey 2006-2007, Govt. of India, Ministry of H.I. and P.E.*
- *Internet*

Space

Today's military planners see Space as the 'high ground' a vital part of the military equation - In Future Wars by Frank Barnaby.

Background

37. Space is the next frontier of conquest by nations having a credible defence capability. Gaining an edge in capability over the adversary has been a constant struggle between nations since time immemorial. In modern warfare too, domination of space is a culmination of the struggle to gain the strategic and tactical high ground, at least till capability is developed to go beyond into outer space. The hearts of the revolution in military affairs (RMA) are space based technologies. Superiority in space dominates the air space and the technical capability of all the ground forces including the functioning of virtually all non military apparatus of a nation. Space dominates all methods of communication and therefore the Internet, GPS, intelligence, imagery, surveillance, commerce, transportation etc. It is like God sitting above.

38. The Gulf war of 1991 witnessed the use of GPS based on satellites and engaging multi targets simultaneously. It also provided real time information, imagery, meteorology, Communication & surveillance. It is also called the first space war The USA used space capabilities to a large extent and this also came to be known as the first space war . The gulf wars and the war in Afghanistan have proved the capability of net centric war (NCW) and its dependence on communication and satellite technology thus optimizing information technology. In a satellite based NCW environment there is a complete meshing of communication horizontally and vertically across land, Sea, Air and Space. It is also integrated through the medium of cyber world. This communication enables access to huge amount of data that is processed into information and made available at the correct level for decision making.At one end information is collected by sensors and after processing at the other end is the shooter.There is bound to be a time lag in this passage of information and action taken thereon. This is the Paradox on one end is a mass of information and on the other shrinking time frame.

39. Space is increasingly being used by large number of countries for their development and communication needs in case they do not have the technology they do so in cooperation with other countries or by outright purchase. Dominance of space therefore is emerging as the strategic high ground of the future.There are serious concerns regarding protecting of space assets from adversaries. This is resolved by two methods first by hardening the satellite but nothing more can be done as they have a predictable orbit and are an easy

target. The other method is to have of fensive capability to protect it self or have hunter satellites to protect own asset s. The second option naturally will st art an arms race in space and is being avoided, as this also makes own satellites vulnerable. The dominant power in space is America followed by the Chinese. The Russians do not have the resources at their command and are maintaining less than 100 satellites that they have as a minimum deterrence. Therefore in the Indian context the American and Chinese st atus is more relevant. The study is primarily concerned with the military use of space.

Military Uses of Satellites

40. the following are the military applications:-

- Imagery – to identify targets to detect underground nuclear detonations.
- Telecommunication – enabling exchange of information between troops deployed in all echelons
- Signal intelligence – detect and capture communications, conversations and broadcasting signals
- Navigation – to assist in accurate navigation of vehicles on ground, sea and air and pinpoint targets for precision munitions.
- Early warning – by use of Infra-red sensors to spot missile launches by detecting the infrared signals emitted.
- Metrology - collect weather data and accurately predict weather conditions

Global Positioning System

41. The Global positioning System is a constellation of satellites in geostationary position in outer space used for accurate location of position on earth. It has wide application in commercial and defence sectors. Disabling this system will disrupt normal life in the modern world.

USA Space Program

42. The space program of USA started in the fifties and gathered momentum after launch of the sputnik by USSR. Current sp ace policy of USA is influenced by the Rumsfeld " Report of the commission to assess United States National security space management organisation." Rumsfeld has stated that this report was not primarily about space weapons, but about how best to organise and manage national security space efforts."

43. As per SIPRI 2008 the US Commission to assess US national security sp ace management and organization, stated that the USA interests are to promote the peaceful use of space and to use space to support US domestic , economic, diplomatic and national security objectives. The attempt to keep weapon out of space while exploiting it for military purpose (without weapon and only for communication, Intelligence and Guidance) suggest a desire to develop and deploy the means to deter hostile acts directed against USA and defend US space assets.

44. The above inter-alia also means continued America dominance and preventing use of space to others, unacceptable to most other countries.

45. The American space program started in the fifties but the salient change that took place was on 4 July 1982 when Ronald Regan announced the National security decision directive number 42 and the Strategic defence Initiative (SDI) also known as the Star war speech. The broad outline of the policy were

(a) strengthen the security of the United States;

(b) maintain United States space leadership;

(c) obtain economic and scientific benefits through the exploitation of space;

(d) expand United States private-sector investment and involvement in civil space and space-related activities;

(e) promote international cooperative activities that are in the national interest; and

(f) cooperate with other nations in maintaining the freedom of space for all activities that enhance the security and welfare of mankind.

This was superceeded by the policy announced by President George W Bush on 19 Sept 1996 as follows:-

> *"(1) for over three decades, the United States has led the world in the exploration and use of outer space. Our achievements in space have inspired a generation of Americans and people throughout the world. We will maintain this leadership role by supporting a strong, stable and balanced national space program that serves our goals in national security , foreign policy, economic growth, environmental stewardship and scientific and technical excellence.Access to and use of space is central for preserving peace and protecting U.S. national security as well as civil and commercial interests. The United States will pursue greater levels of partnership and cooperation in national and international space activities and work with other nations to ensure the continued exploration and use of outer space for peaceful purposes.*

(2) The goals of the U.S. space program are to:

(a) Enhance knowledge of the Earth, the solar system and the universe through human and robotic exploration;

(b) Strengthen and maintain the national security of the United States;

(c) Enhance the economic competitiveness, and scientific and technical capabilities of the United States;

(d) Encourage State, local and private sector investment in, and use of, space technologies;

(e) Promote international cooperation to further U.S. domestic, national security, and foreign policies. "

47. With this as a backdrop there is a fear amongst other nations especially china that the missile defence program being developed by USA purportedly for defence against missiles attack from rogue states is also directed against it and it would neutralise its nuclear deterrence. Therefore to maintain its parity it is developing means to protect its assets in space including by developing Anti Satellite capability that it demonstrated in Jan 2007, it is also developing its GPS and military satellites. Like wise the European countries and Russia also have their concerns and are developing their own GPS Galileo. With these developments America fears a space 'Pearl Harbour". The Rumsfeld commission has selectively highlighted provisions in each of its international treaties, searching for arguments to permit deployment of ofensive and active defensive systems rather then recommending the strengthening of the provisions to prevent weaponisation of space. Perhaps the commission felt that this would give America a legal route to weaponise space to its advantage.

48. The National security Strategy announced by President Bush on 17 Sept 2007 is as follows:-

" *The gravest danger our Nation faces lies at the crossroads of radicalism and technology Our enemies have openly declared that they are seeking weapons of mass destruction, and evidence indicates that they are doing so with determination. The United States will not allow these efforts to succeed.* ***1. We will build defenses against ballistic missiles and other means of delivery2. We will cooperate with other nations to denycontain, and curtail our enemies'efforts to acquire dangerous technologies.And, as a matter of common sense and self-defense, 3America will act against such emerging threats before they are fully formed.***

THE WHITE HOUSE
September 17, 2002"

49. The Vision of the US Space command responsible to implement the strategy is: -

"Dominating the sp ace dimension of milit ary operations, to protect US interest s And investments, integrating Space forces into war fighting capabilities across The full spectrum of conflict."

It intends to achieve the same through the following:-

1. Control of Space is the ability to assure access to Space , freedom and operations within the space medium and an ability to deny others the use of space, if required by
   - Real time space surveillance
   - Timely and responsible space lift.
   - Enhanced protection to military and commercial assets.
   - Robust negotiation system
2. Global engagement is the application of precision force from, to, and through space. This will require national missile defence (NMD) through the following
   - Non intrusive global surveillance.
   - Key to NMD
   - Enhanced C2
   - Space based strike weapon.
3. Full force projection is integration of sp ace forces and sp ace derived information with land, sea, Air forces and their information. Getting the right people at the right time using.
   - Enhanced sensor- to shooter time.
   - Common protocols, communication standards, and fused data bases.
   - Precise modeling and simulation
   - One-stop shop for space support.
4. Global partnership augments military space capabilities through the leveraging of civil, commercial and international space systems. Growth of US non-military space

systems provides the opportunity for the US to gain increased battle space, awareness and information connectivity in a cost effective manner. It offloads pressure from military functions to civil/ commercial partners by

- Sharing of space based information.
- Influencing space system design.
- Satellite sharing.
- Space system architecture to facilitate rapid flow of information.
- International standardisation.

50. The response may not be lagging wherein China may extend the limits of its aerospace sovereignty and find an equally legal reason to use offensive means to neutralise the adversary satellite. Even if the US were to enforce its dominance by export control, treaties or other means the dual nature of the use of satellite technology still leaves a risk that benign development will not be used maliciously (SIPRI 2008).

51. On 14 Feb 2008 USA launched a SM-3-Anti ballistic missile to knock down a spy satellite USA-193 reportedly since it contained toxic hydrazine fuel. It is however believed to be a demonstration in response to the Chinese similar operation in 2007.

52. **Lacrosse** and **Onyx** are the code names for the United States National Reconnaissance office terrestrial radar imaging reconnaissance satellite. While not officially confirmed by anybody in the government, there is widespread evidence to confirm its existence, including one NASA website: (httt://nssdc.gsfc.nasa.gov/amc/ spacecraftDisplay.do?id=1988-106B).

53. Lacrosse uses Synthetic aperture Radar as its prime imaging instrument. It is able to see through cloud cover and also has some ability to penetrate soil, though there have been more powerful instruments deployed in space for this specific purpose. Early versions are believed to have used theTracking and Data Relay Satellite System (TDRSS) to relay imagery to a ground station at White sands New Mexico.There are some indications that other relay satellites may now be available for use with Lacrosse.The name “Lacrosse” is used to refer to all variant s, while “Onyx” is sometimes used to refer to the three newer units.

Five Lacrosse spacecraft have been launched, with four currently in orbit. Lacrosse 5 appears to differ somewhat from the previous four satellites. As determined by amateur observers, there are subtle differences in its orbit and its color is somewhat whiter than the distinct red-orange tint of the earlier ones. Most strikingly, it sometimes fades from brightness to invisibility within the space of a few seconds while still in full sunlight. (The amateurs call this rapid fade its "disappearing trick".) (Source Internet).

Chinese Space Program

54. The Chinese Space Program started as early as 1957, emerging from its perception of threat from the USA during the Korean War. It was adapted using the soviet technology by Dr Chien Hsu-shen an engineer at the California institute of technology who was expelled from the USA as part of the anti-communist purge in 1955. It assumed greater importance after support from the USSR to China on military technology stopped. For the Chinese leadership from Mao onwards Space & military technology along with nuclear technology was the main line of defence. This was due to weakness in conventional defence capability which the leadership wished to cover. Mao steered an independent foreign policy for China which would not have been possible in the absence of total self reliance in defence capability. It precluded what Mao termed as 'Bullying' by other Nations. Chinese thoughts on war are strongly tampered by Sun Tzu where the first tenet is 'know the enemy and know yourself" and this has perhaps influenced its policy to understand its conventional weakness and its efforts to overcome this by gaining asymmetric parity by developing its nuclear weapon, cyber war and space capabilities.

55. Chinese concerns are further heightened by the American training exercises which portray a nuclear country (read China) threatening a small neighbours (read Taiwan) where the US is drawn into the conflict and there is a clash of space assets. Further US reportedly are constantly tracking the over 800 satellites currently in orbit including military satellites. This reports suggests that the ability to monitor and track other satellites also provide targeting data to America for its anti satellite weapons.' (Jamestown Foundation report on China dated 8 Apr 2008).

56. China feels threatened in the emerging world order of American dominance. It knows that it cannot match America in conventional terms and is therefore vigorously pursuing a path to gain parity by asymmetric means to safe guard its interests as a growing economic power. It has therefore since early sixties developed a credible nuclear capability initiated under Mao Tse Tung. The next step was to perfect the delivery system. With the current development of communication and information technology it found its nuclear deterrence declining especially due to emergence of satellite enabled technology demonstrated during the gulf wars. China therefore feels that it would loose out if the deterrence were not continued through space, thus its vigorous implementation of the program which, as is widely known, is more military than civilian. Once China deploys its assets in space it

would develop the same vulnerability that the US feels today and hence the attempt s to develop anti satellite means (covert at present). If there is an open arms race in outer space it would emanate due to clash of interest between these two countries.

57. China totally shifted its attention to developing its space capabilities and its defence policy published as a white p aper stressed the use of outer space for developing communication, navigation and reconnaissance. It shifed its opposition to militarization of outer sp ace to weaponisation of S pace. The 2006 white p aper omitted even this. The preeminence of USA to have weapons in space neutralizes the deterrent of China nuclear ICBM's. China is therefore concerned on this issue. This resulted in the knocking down of a satellite by china in Jan 2007. The fears of china are reinforced by USA shooting down a satellite using an SM-3 interceptor missile (China *brief – Jamestown foundation 14 Apr 2008)*.

58. The current cap ability of China is based on success achieved through const ant development of technology over the years. Its success depended upon:

- Leadership at the highest level provided strong and consistent support to the program especially for military development
- Policy of self reliance enabled China to develop technology in-house while it welcomed technology from wherever possible.
- Space development was accompanied by building a space Industry.
- In early years space and military technology was common.

59. In 1978 the Chinese leadership called on the defence sector to contribute towards the growth of the economy This caused a shift in the defence Industry to shif focus towards commercialization in order to subsidize the defence expenditure and earn revenue(Yanping Chen in 'China's Space Interest & Mil tech control') .The successful launch of Chang'e-1 into a lunar orbit heralds a deep space capability. Has achieved breakthrough in the following fields (Jiefangjun Bao, 25 Oct 2007)

- Orbit design.
- Directional attitude control.
- Adaptability to the peculiarity of the space environment.
- Response to the influence of the lunar eclipse.
- Long range telemetry, tracking and control.

- Reverse engineering of scientific data.
- Spin-off benefits in the use of technology for civil use.

60. "In October 2003 it became the third country in the world to put a man in space using its own rocket. In Oct 2005 it sent two Taikonauts (Astronauts) into space abroad Shenzhou VII spacecraft. On 24 Oct 2007 the Chinese launched the Chang'e -1 spacecraft in lunar orbit. These are part of 'Project 921' program. In Sept 2008 Chinese Taiknauts carried out a space walk. China has a program to have a vehicle orbiting the lunar surface by 2015 to collect soil samples for study. In Apr 2008 China launched its first data relay and tracking satellite giving it real-time military intelligence capability." (Jamestown foundation)

61. China is aggressively promoting the sale of space products to other countries both for political and commercial reasons. There are three independent companies which market space products as part of Chinese policy of internal competition. China thus intends to cover its conventional weakness by a forward leap in employing technology at a much lower cost. Its foreign policy towards international initiatives is therefore, tampered with a backdrop of complex security and commercial interests. In case of any international agreement that affects its defence or commercial interests China is not likely to be a signatory unless it gains or compensated for it elsewhere. It feels that the MCTR may not be successful for four reasons:

1. MTR is the prevailing views of USA to protect its interests at the expense of third world countries.
2. The Pact may not attract many countries because almost 100 countries are engaged in space technology R&D and 20 countries have their own space programs. Also outer space belongs to humanity and MCTR is a means to monopolize it.
3. Enforcing the MCTR would be difficult since it is a Law, a Treaty or an Agreement.
4. Lastly while it restricts non American arms sales in the world it seeks to promote those of America. These double standards are difficult to accept.

62. As per US China Economic and Security Review 2007, their assessment the Chinese space program is essentially a military program to achieve a leap forward. China considers the American dependence on space as its soft ribs and strategic weakness and Chinese investments in this field can provide it a potentially asymmetric capability. Chinas space program covers military, intelligence, and reconnaissance, earth monitoring, R&D, scientific exploration, communication, media, and military command & control. The space program in China has additional benefits of raising its international prestige, unifying the country

internally and imparting it political stability. It contributes to the economic development and enhances the capability of the PLA. Chinese Anti satellite capability enhances its offensive ability by dominating the battle space. Besides this it has commercial benefits of earning revenue by sale of space products. Chinese intentions are also apparent from its earlier shift to oppose weaponisation of outer space to being silent about it in the white paper of 2006 and exploiting the legal loop holes to advance its own program.

Chinese Space Assets

63. Following are the Chinese space assets abroad.

- **Tracking stations abroad** - In Pakistan, Kenya, Kiribati and Namibia
- **Communication satellite** – APStar, Asiasat, Sinosat, Chinasat
- **Imagery, Remote sensing & Synthetic Radar signals** – Zinyuan, Cbers-2, Haiyin, Jianbing, Huanjing.
- **GPS** – Compass system for navigation. China has 5 satellites over east China and west pacific. Plans 30 more in the future.
- **Anti-satellite** – developed anti satellite of co-orbital direct attack weapon, direct energy weapon and satellite Jamming capability being developed.

Future Space Developments

64. Future space developments are categorized as:-

- Developing heavy lift capable engines for future space stations.
- Performing in orbit docking, to assist in building space stations.
- Develop a Lunar exploration Rover by 2015.
- High resolution earth observation satellite for civilian and military use
- Develop ground relay telemetry stations for increased access to satellites.
- Improve Compass navigation satellite system for worldwide coverage.

Launch geostationary orbit communication satellite for uninterrupted communication.

65. China tested its anti satellite capability in Jan 2007 by destroying a weather satellite FY-1C 865 Km above earth using a SC-19 ANTI SATELLITE. The missile was launched from a mobile launcher and hit the target head-on destroying it. This drew a lot of international

criticism. China is also cooperating with countries like Argentina, Canada, Malaysia, Pakistan, Singapore, and Ukraine on space cooperation and has exchange programs with Algeria, Chile, Germany, Italy, Japan, and Peru. These are with a view to expand its technology base and also to further its commercial interests in future. China is closely watching the American space program and keeping in view its heavy dependence on satellites it feels it is important to have a capability to Jam or disrupt the enemies satellite this would give the PLA some chance of success against the vastly technologically superior opponent.

66. **Chinese Military Satellites -** Fanhui Shi Weixing means “recoverable satellite” is a series of Chinese recoverable satellite. The satellites were designed for both military and civilian observation needs, with a total of 26 flights. The first Flight was FSW-0 on 1ST June 1969 and the last SJ-8 on 9th Sept 2006. Four models of the satellites were introduced: FSW-0, FSW-1, FSW-2, and the most modern being FSW-3. All satellites were put in orbit using the long march rockets, most of them by the CZ-2C type.

Indian Military Space Program

“Networking and assimilation of space, both interdependent, are already the way forward and can tilt the balance considerably. A quiet ‘space race’ is the current reality and weaponisation, a distinct possibility.” – Air Marshal FH Major, Chief of Air Staff.

67. The Indian space program started in 1962 with the formation Indian national committee for Space research and has since reached a stage comparable with the best in the world; Today India is the fourth country in the world to have sent a mission to the moon. The important mile stone in Indian space technology are:-

- 19 Apr 1975 - Aryabhatta the first Indian designed and fabricated Satellite Launched in space with Soviet help.
- 1979 & 1981 Bhaskara 1 & 2 observation Satellites respectively launched.
- 1979 – 81 Rohini series of Satellites using Indian launch Vehicles Successfully placed in low earth orbit in the second attempt.
- 1981 Apple experimental communication launched on Ariane vehicle 1981.
- 1980 The INSAT series communication satellites launched .
- 2001 GSVL Geo synchronous launch vehicle put in orbit.

68. Military space program is virtually negligible. Though the Indian space and Military

community fully understand the importance of a military capability in space yet no headway has been made primarily due to internal problems like considerations of Turf control, and lack of Political conviction. The Indian Integrated guided missile development program (IGMDP) has achieved mixed success. Its biggest advantage has been the development of the solid rocket motor, control, guidance and navigation technology developed for the SLV-3 and ASLV boosters have been adapted for the Agni range of ICBM. It will also benefit from the PSLV (Polar satellite launch vehicle) and GSLV (Geosynchronous Satellite Launch Vehicle) technologies.

69. The INSAT series of eight satellites provides the largest domestic communication system in the Asia-Pacific region operating 130 telephone exchanges and Television services.

70. The IRS (TES) launched in 2001 enables imagery of up to one meter resolution and provide valuable data on natural resource survey and management. It also has military utility. The paradox is that the Indian army has to approach Israel for space imagery despite having the largest number of remote sensing satellites. In 2003 ISRO launched the CARTOSAT -1 and the follow up CARTOSAT-2 satellites to enable the military to get geographic information systems (GIS).The GAGAN – Global positioning system has been at the request of the Ministry of civil aviation.

71. On 22 Oct 08 India successfully launched it mission Chandrayaan-I an unmanned space craft to the moon. It is India' s first attempt to the moon and has been carried out flawlessly in the first go. Further the cost is just $ 400 million, a feat by itself. The Payload is:-

- Terrain mapping camera to prepare a 3D atlas of the moon
- Hyper spectral imager will obtain spectroscopic data for mapping minerals on the moon
- High energy X-ray spectrometer will detect radioactivity signals and identify chemical compositions.
- Lunar laser ranging instrument will help keep the spacecraft in proper orientation above the moon
- Moon impact probe will detach from the craft and crash land on the moon surface, putting the Indian flag on the moon.
- One of the most important achievements from the commercial angle is that it is

carrying six international payloads reinforcing India's capability in the Space products market.

Importance of Space Programs for India

72. The Indian space program is for the purposes of satellite based applications like communication, surveillance, remote sensing and global positioning and meteorology. Therefore the Indian space program must be viewed in light of international development in outer space. USA is the leader, dominating space and would like to maintain this lead. It has abrogated the ABM treaty in June 2002 and is developing space weapons capability though not overtly else this would start an arms race in space. Russia has neither the capacity nor the resources to deploy satellites in space on the lines of the erstwhile USSR, it is therefore maintaining a minimum deterrent. China is the next contender and feels that its biggest threat is from USA and is therefore modernizing to neutralize that threat. It is fast developing its missile capability towards this end. In Jan 2007 it knocked down one of its own obsolete satellite successfully purely to test and demonstrate its anti satellite capability. It is also reportedly developing parasite satellites that are small and attach themselves to another satellite and are activated on command. All these are a pointer to china's serious concern to overcome its conventional weakness and gain parity by other means. The Indian concern should therefore be that China's military capability can be directed against India. India therefore, whether it likes it or not, is required to factor these concerns in its planning. As per press reports of 9 Jan 2008 USA would like to partner India in its international missile defence program using the PAC-3 system however there is no official conferment.

73. Further India's ambition as a regional power along with a policy of following an independent foreign policy devolves upon it to develop independent assets. The path will have to be totally indigenous. Therefore it must have its own GPS satellites plus a network of communication and remote sensing satellites. It must have capability of early warning and harden its satellites as a passive measure against attack. There is a need to establish more than one network as a standby measure. Reserve satellites will have to be kept for immediate launch in case of a satellite being disabled.

74. An indigenous effort like the Chandrayaan mission needs to be developed further to develop technology and reduce costs. There are strong arguments against the Chandrayaan mission that first India cannot afford it and second that while USA & Russia have already been to the moon they have not taken steps to further it. The Indian mission is not purely for prestige, but has commercial interests in it. It also has interest not to lag behind in the emerging technology which will put it at a strategic disadvantage later. Whatever the

argument against it, weaponisation of sp ace is on and we cannot ignore it. While India need not attempt parity with USA yet it must draw lessons from history when it missed the industrial revolution. Technology and infrastructure to develop a minimum deterrent must be put in place in pure national interests while espousing the cause of non weaponisation of outer space (it implies that all countries must renounce it including USA).

75. The Indian sp ace program st arted in 1962 with the formation of 'Indian national committee for space Research' (INCOSPAR). This was upgraded to ISRO on 14Aug 1969. The mission profile of ISRO for the 2006-10 AND 2010-13 are given below :-

ISRO Mission profile 2006 to 2010				
Mission	**2006-7**	**2007-8**	**2008-9**	**2009-10**
Earth Observation	CARTOSAT-2	COMMERCIAL LAUNCHES OF AGILE & TECSAR	TWSAT CARTOSAT-2A, 2B OCEANSAT-2	RESOURCE-2 MEGHA-TROPIQUES RISAT-1 INSAT-3D
Satellite Communication & Navigation	INSAT-4C INSAT-4B	INSAT-4CR	GSAT-4 GSAT-5 GLONASS N1	GSAT-6,8,9 IRNSS-1,2 GLONASS N2
Space Science & Environment	SRE		CHANDRAYAA-N-1	ASTROSAT SRE-2
Launch Vehicles	F2 C7	PSLV-C8&10 GSLV-F4	PSLV GSLV F5 D3	PSLV GSLV F6 F9

Source: Department of space India annual report 2007-08

ISRO Mission profile 2010 to 2013			
Mission	**2010-11**	**2011-12**	**2012-13**
Earth Observation	TES-HY3 ALTIKA ARGOS GEO-HR	RESOURCAT-3 DMSAR-1 CARTOSAT-3	RISAT-3 OCEANSAT-3
Satellite Communication & Navigation	ACTS-1 GSAT-7,10 IRNSS-3,4	GSAT-11,12,13 IRNSS-5 IRNSS-6	ACTS-2 GSAT-14 GSAT-15 IRNSS-7
Space Science & Environment	CHANDRAYAAN-2 I-STAG	ADITY-1 SENSE-P SENSE-E	
Launch Vehicles	PSLV GSLV	PSLV GSLV	

References

1. *New technology and the arms race – H Macmillan 1989.*
2. *High technology and reducing the risk of war – New York academy of science 1986.*
3. *Science and technology NDC – 1994.*
4. *US congress hearings committee on science& Space technology- Washington 1991.*
5. *Report on Indian space program 1996.*
6. *Illustrated encyclopedia of space technology.*
7. *Space powers Interests – edited by Peter Hayes IS BN 0-8133-8879-1(HC)*
8. *Thomas G Mahnken & Janne Nolan in 'space powers Interests".*
9. *SIPRI Yearbook 2008.*
10. *10. Sqn leader KK Nair in Space the frontier of modern defence.*

Missiles

76. Modern warfare is unthinkable without missiles. Missiles have been developed to operate from land, sea, air and underwater against various targets. These are being used in land warfare, air warfare and sea warfare including a combination of all of the above. Thus we have witnessed missiles launched from the sea targeting the Al-queida elements in Afghanistan using satellite based tracking & guidance. It only reiterates the tremendous versatility in the use of missiles as a precision weapon system in modern warfare where collateral damage is unacceptable in operations that are transparent to the world at large.

77. Missiles are used in the defensive & of fensive role both as a t actical and Strategic weapon. In the tactical and Strategic areas the use of missiles difers in their range and or their payload. Essentially, at the heart of missile technology are the main components i.e. the airframe, engine using dif ferent propulsion systems, the guidance and warhead or payload for which the missile is intended to be used. The technology for each individual component of the missile is att ainable and India is competent to develop and deploy the same, however its integration with the detection of targets and launch needs to be constantly upgraded to keep p ace with increase in technology of enhancing speed, reducing radar signature by stealth technology and low / high flying missiles that are difficult to detect. It will be apparent that the difference in avionics of offensive and defensive missiles is little, except its deployment integrated with the tracking & guidance system in the of fensive or defensive role.

78. The areas where innovative use of missiles is emerging are:-

- As a standoff weapon system on surface and in the air.
- As a long distance low altitude precise weapon system in surface warfare.
- In the missile defence (MD) role.

79. In all of the above roles it is part of an integrated system where tracking of the target is achieved using sophisticated technology incorporating radars, satellites and interpretation of tell t ale electronic and infra-red signatures of the t arget. Once identified positively , a defensive missile is launched and locked on to the hostile target to accomplish its mission of its destruction.

USA – Missile Program

80. The genesis of missile can be traced back to 1950s' when ICBM's armed with weapon of mass destruction had started to appear, since measures for effective defence could not

be taken due to lack of technology . It was simpler to make missile armed with WMD to create deterrence. This resulted in a race culminating into a st ate of mutually assured destruction (MAD) level of deterrence. In 1972 the deterrence was institutionalized by the anti ballistic missile treaty (ABM), placing restrictions on the superpowers to develop and deploy missiles. A further development was the caps placed on the number of weapon on each superpower under the Strategic arms limitation talks (SALT) agreement of 1972 and 1979. But the peace was on gunpoint and uneasy. In 1981 the Missile control technology Regime (MCTR) was formed to stop proliferation of missile technology it has a membership of 22 countries, but has not succeeded in it s objectives. Details of the MCTR are given below:-

If the scenario on missile technology is as discussed above, it is but normal for those having the technology to raise the barrier of entry to others thereby maintaining their monopoly and its concomitant advantages. The missile control technology regime was formed in 1981 with the lead being taken by the USA, ostensibly to stop proliferation of the technology to irresponsible st ates. It has a membership of 22 countries. Restrictions were imposed to limit the spread of missiles capable of delivering a payload of 500 Kg at a dist ance of 300 Km. It s aim is to deny technology of the complete delivery systems including complete rocket systems such as ballistic missile including rocket stages, re-entry vehicle, guidance system, thrust, vector controls, warhead-safing, Firing, Arming. These are the category I systems. In category II systems are the dual use missile components which, is to be discussed on a case by case basis.

The limits of the MCTR are as follows:-

- A cartel to limit the supply of technology will provide a strong incentive to cheat (witness the case of Nuclear technology).
- Restriction on the source of supplies will cause a new source to be established outside the cartel. Countries like China and Korea will have a strong incentive to proliferate the technology.
- The standards of enforcement are different among the MCTR members. The association is also voluntary and lacks institutional arrangements.
- Since the membership is limited some countries have formed their own rules. Russia has its own policy on the subject. China has agreed to bide by the rules but merrily continues to proliferate and export technology.

(*Thomas G Mahnken and Janne Nolan - in* Space powers interests)

It was in 1983 that the US sought permanent peace by surpassing MAD and following an alternate route to weapon deployment. President Ronald Regan made the historic 'S tar war' speech authorizing the development and deployment of a ballistic missile defence (BMD) of the US on the grounds that it was immoral for the population to live under fear of destruction. The aim was to adopt an edge over the Nuclear weapon deterrence. He proposed the Strategic defence initiative (SDI) which entailed a missile defence constituting anti ballistic missile deployed to protect the US along with space based surveillance and detection system to intercept enemy missiles. Laser based weapon in sp ace were to be deployed to neutralize the incoming missiles.

81. The Soviets had to respond to this by utilizing immense resources that they could not afford to counter the SDI to gain parity if not a superior edge for which they did not have the technology. They negotiated the eventual end of the cold war leading to the end of USSR. Whether this alone caused or contributed to the breakup of USSR or not, the fact is that Credit must be given to Ronald Regan, who in true SunTsu style won without firing a shot. The US itself did not have the technology at that point of time to deploy an effective BMD, and still do not have.Technically it has been demonstrated to be possible but the proposed impregnable shield over the country is not at present possible. In the sense it is not fail proof. Consequent to the breakup of the USSR the reason for the SDI also vanished and the program was abandoned.

82. In 1998 the Rumsfeld commission found strong backing in the US congress to contain threat not from a major adversary but from Rogue states consisting of small under developed countries having regional ambitions. These states pose a threat to the US who was likely to intervene in regional affairs to maintain the status quo. Meaning America would like to have states that toe their line else they are perceived a threat to US long term interest s whether oil, other resources or commerce. The subject has other political and economic connotations and is complex. The idea of building a MD was also not accepted by the European allies, China and Russia. President Clinton deferred decision on it and it was George Bush who pushed for this concept. In Dec 2001 Bush announced a unilateral withdrawal of the US from the ABM treaty. This freed the US from any encumbrance at a time when opposition was weak or non existent, paving the way for the USAto be the sole power in the world. *(Bertel Heurlin, Sten Rynning & Kristian Soby Kristensen in 'Missile defence international, Regional and national implications" – Routledge 2005)*

83. The recommendations of the Rumsfeld commission resulted in the enacting of the American national missile defence Act-1999 as follows:-

" The National missile defense act dated 21 Jan 1999*To deploy as soon as is technologically* *possible an effective National Missile defence system capable of defending the territory of the united states against limited missile attack."*

84. Next year in Jan 2002 the Secretary of state Donald Rumsfled USA, announced the redesignation of the Ballistic Missile Defense Organization (BMDO) as the Missile Defense Agency (MDA). This is consistent with the president's emphasis on missile defense as per the 'The National 1999 Missile defense Act", the secretary also provided direction necessary to meet the top four priorities for the United States in this important mission area. These are:-

- To defend the United States, deployed forces, allies and friends from ballistic missile attack.
- To employ a Ballistic Missile Defense System (BMDS) that layers defenses to intercept missiles in all phases of their flight (i.e. boost, midcourse, and terminal) against all ranges of threats.
- To enable the Services to field elements of the overall BMDS as soon as practicable.
- To develop and test technologies, use prototype and test assets to provide early capability, if necessary, and improve the effectiveness of deployed capability by inserting new technologies as they become available or when the threat warrants an accelerated capability.

85. Elevating BMDO to agency status recognizes the national priority and mission emphasis on missile defense. Full and cooperative efforts of the military Services, Joint Staff and defense agencies to the Program have been directed. The overall objectives for missile aim at establishing a single program to develop an integrated missile defense system, assigning the best and brightest people to this work and applying a capability-based requirements process for missile defense. This is supported by the budgetary support for the program as follows:-

Year	2003	2004	2005	2006	2007	2008	2008 – 13 *
Funding	6714	7674	9169	7695	9388	8899	56666

(US Deptt of defense fiscal year 2008 budget estimates in Mn $) * Projected

The important part of the policy is that to accelerate the deployment an evolutionary approach has been adopted. Even this would add to the planning of an adversary as an extra element to cater for in his plans, thereby achieving extra deterrence for the US. *Implying that the USA will certainly go ahead with the Missile defense (MD) program.*
www.defenselink.mil/release.aspx?releaseid=3203>

86. While the technology has been demonstrated its implementation is another story, yet as a matter of policy the USA continues to deploy it knowing it is not fool proof. The argument forwarded is that spiral development would continue as technology develops. The deployment of an embryonic system is a fait accompli to be used as a legal shield for future weapon proliferation. The aim is to retain or prolong the American leadership as a dominant world power. Wilhelm Agrell in his essay "Pre-empt, Balance or Intercept" – Routledge, has convincingly stated

"The nearly two decades old issue of the interpretation of the ABM treaty was finally pushed aside by the George W Bush administration in Dec 2001...... At that time, the build-up had started for the first preventive war explicitly legitimized with the need to deprive a state once and for all, of the capability, to use or develop WMD and their delivery systems."

87. As per SIPRI, in Sept 2007 the US successfully carried out test of a ground based interceptor (GBI) missile. It also included the ability of the upgraded early warning Radar at Beale air force station at California. The target missile was also successfully tracked from sea by an Aegis cruiser using AN/APY-1 Radar. It also tested the SBX radar that acquired and tracked the target missile. On 6 Nov 2007 two target missiles fired from Hawaii were successfully intercepted by the ship 'Lake Erie' in the Pacific Ocean. While the test flights were successful they do not assure full defence in actual conditions when counter measures will be adopted by the hostile missiles. USA has entered into a Missile development programs simultaneously with other countries. These are:-

88. **GMD** – ground based midcourse defence system. The US intends to deploy missiles in Poland and the Czech Republic to counter threats to Europe from Iran's emerging long range ballistic missile. The first base would be in Koszalin in Poland and the second in the Brdy district of Czech. The interceptors would be silo based and would be a two stage variant of the Ground based interceptor (GBI). They will have a greater acceleration but a shorter range and better suited to engage missiles aimed at Europe, since the three stage interceptor minimum flight range and time is too long to engage missiles launched from Iran. The missiles based in Czech would have a X-Band radar called the European midcourse radar (EMR). The narrow beam high resolution radar will allow discrimination of target clusters and identify missiles from other counter measures. The EMR will also enable precision tracking and guidance information called '**cueing data**' to the interceptor missile thus enhancing its capability. A X – based radar may also be deployed closer to Iran in Turkey.

89. **NATO** – A theatre missile defence system called 'Active layered theater ballistic missile defence' (ALTBMD) program to protect its deployed forces within and outside its

territory against short and medium range ballistic missiles. It is a multi layered system consisting of Low- and High altitude defence, communications, command and control systems, early warning radar and various interceptors. This will integrate into a single NATO command and control network sensors and interceptors. These will be provided by the members like Franco-Italian system SAMP/T and the US Patriot anti-missile system. The ALTBMD is scheduled to be deployed by 2016.

90. **MEADS** – The Medium extended air defence system (MEADS) by a German- Italian – US air defence programs. It consist s of a lightweight launcher, 360 degree fire control and surveillance system interoperable with the NATO forces. It will use the US army PAC-3 interceptor missile compatible with MEADS. It will be in service by 2014.

91. **Japan** – the Japanese cabinet approved the equipment and technology exchange and cooperation on ballistic missile defence with USA. Joint production of the US Standard missile-3 interceptor Patriot Advance Capability (PAC-3). Japan is also developing a high altitude exo atmospheric anti-missile capability that will consist of the Japan maritime self defence force (JMSDF) destroyers equipped with the Aegis BMD radar and weapon system and SM-3 missiles. These missiles have on board high resolution Radar, computer processor and 180 thrusters for fine tuning of the final homing in to the target. These initiatives of Japan with USA are a result of the long range missiles developed by North Korea.

92. **Israel** – The US - Israel Arrow weapon system (A WS) is the technologically most mature. It consists of the Arrow 2 interceptor equipped with infrared and active radar sensors and blast fragmentation warhead. The US will also fund the further development of the Arrow system. (SIPRI Yearbook 2008).

93. What emerges is that the USA has decided to deploy the Missile defence system and also to expand it to be deployed closer to possible origin of hostile missiles in cooperation with its allies in Europe, Israel and Japan. USA is also keen to extend the missile defence program to other countries with a view to achieve its objective by engagement, therefore cooperation in missile defence technology is one of the clause in the nuclear cooperation strategic pact signed between India & USA in 2008. It is well understood that the MD is not perfect but till the technology is developed this will form another line of defence. This also precludes the placing of any restrictions on the US in future by any international treaty.

Summary of Principal US Missile Defence Programs (Dec 2007)

Details as available from SIPRI -2007 are given below:-

Srl No	Program	System	Status
1	Patriot Advanced Capability PAC-3	Land based , Air-Transportable, extended range, Phased array Radar & engagement control station	Most advanced. Total 712 to be in service by 2008
2	Terminal High Altitude are Defence (THAAD)	Truck mounted launcher-hit to kill interceptor missile. X band Radar, Battle management & control system (BMC2)	Resumed test flights after major design changes in 2005
3	Ground based Mid-course defence(GMD)	Ground based interceptor, long range, Land & sea based tracking	GBI & GMD Radar network successfully used in test 2007
4	Aegis Ballistic Missile defence	long Aegis Ships equipped with AN/SPY-1 Radar. Long range surveillance and track, Hit to kill interceptor.	3 Aegis BMD cruisers and 13 destroyers to be in naval service.
5	Multiple Ballistic Missile Defence	Long range interceptor, carrying 8-20 miniaturized EKVs which can independently track & target multiple warheads & Mid course countermeasures, decoy & reentry vehicles	Control system of payload carrier vehicle tested in 2006, testing of EKVs to begin in 2009
6	Airborne Laser (ABL)	Modified Boeing 747 aircraft, carrying megawatt class chemical oxygen iodine laser. Infrared sensor & target acquisition & tracking laser	System integration problems, first lethality test of laser in 2009
7	Kinetic Energy Interceptor (KEI)	Fast burn high velocity 8interceptor, deployed on m9obile land launcher or sea based platform	Land based booster flight to be tested in 2008 initial operational capability to be tested.
8	Sea based X band Radar (SBX)	High resolution Radar, based on maneuverable offshore platform for acquisition, tracking and discrimination of target missiles.	Completed sea trial in 2007 to be based in Alaska.
9	AN/TPY-2 Radar	Transportable high resolution radar for detecting, tracking & discrimination missile threat designed as part of THAAD	First radar activated in Oct 2006 by US Army units in Japan
10	Space tracking and surveillance system (STSS)	Constellation of low earth orbit satellite designed to detect & track missiles in all phases of flight	Two satellite to be launched in 2008
11	Space based Infrared system-high (SBIRS-High)	USAF procurement plan is for 3 satellite with infrared sensors in highly elliptical orbit to provide early warning of BM launches	Beset by technology delays launch of first satellite scheduled for 2008
12	Upgraded Early warning Radar (UEWR)	Modified early warning radar (EWR) for detection & tracking of post boost & mid-course re-entry vehicle data transmitted to 2 GMD fire control centre in Alaska and Colorado	Upgrade of US ewr at RAF flying dales UK. Upgrade of US EWR at Thule Greenland

Chinese Missile Program

94. As part of its priority defence planning China is developing its missile program concurrently with its space program discussed above. While the claim of US is that its MD program is directed for safety from Rogue states, China views it as a system that has an inherent capability to neutralize its minimalist deterrence of nuclear weapons and aimed to counter its ICBM's. (Peter Dyvad in Missile defence – Routledge 2005) this is reinforced by the following :-

- Location of EW X- band radar located in Alaska form where it can detect not only missile from North Korea but also China.
- US criticism of China's Human rights policy
- America support to Taiwan and China's view of it as a Rogue state.
- US policy of containing the spread of communism.
- Military treaty with Japan, Philippines, Korea south.

95. The Chinese strategic missile program is designed to support its nuclear deterrence. China had proceeded on the route to develop nuclear weapon in spite of international isolation for years so that it could pursue an independent policy . China has developed a minimum deterrence without gaining a parity in quantity with USA and fears that once the US implements its MD program it will neuter this deterrence. While the MD is stated to be against Rogue states it applies equally well against China being in proximity of Korea.The sophistication of Chinese nuclear weapons is far behind that of US, Chinese fears are also based on the large number of missile proposed to be deployed by the US in its MD program. China also does not have an early warning system and this only enhances its vulnerability against a pre emptive attack. Therefore the option open to China would be to increase the number of warheads in the hope that some of them will slip through the MD defence or develop an alternative in space & cyber capability.

96. It was after the Gulf war that focus shifted very strongly on the use of aerospace. The USA used space capabilities to a large extent and this also came to be known as the first space war. China totally shifted its attention to developing its space capabilities and it defence policy published as a white paper stressed the use of outer space for developing communication, navigation and reconnaissance. It shifed its opposition to militarization of outer space to weaponisation of Space. The 2006 white paper omitted even this. The preeminence of USA to have weapons in space neutralizes the deterrent of China nuclear ICBM's. China is therefore concerned on this issue.This resulted in the knocking down of

a satellite by china in Jan 2007. The fears of china are reinforced by USA shooting down a satellite using an SM-3 interceptor missile (China *brief – Jamestown foundation 14 Apr 2008).*

97. China has two series of ballistic missiles the Dong Feng series (DF) and the Long march series (LM) of launcher for satellites. Chinese missiles are classified as follows:-

ChineseClassification	Range (Km)	Comparable classification
Short range (Jingcheng)	< 1000	Short range ballistic missile (SRMB)
Medium range (Zhongcheng)	1000 - 3000	Medium range ballistic missile (MRBM) & Intermediate range ballistic missile (IRBM)
Intermediate range (Zhongyuancheng)	3000 - 4800	IRBM
Long range (Yuancheng)	3000 - 8000	IRBM & Intercontinental ballistic missile (ICBM)
Intercontinental (Zhouji)	> 8000	ICBM

98. In case of any international agreement that affects its defence or commercial interests China is not likely to be a signatory unless it gains or compensated for it elsewhere. It feels that the MCTR may not be successful for four reasons:-

- MTR is the prevailing views of USA to protect its interests at the expense of third world countries.
- The Pact may not attract many countries because almost 100 countries are engaged in space technology R&D and 20 countries have their own sp ace programs. Also outer space belongs to humanity and MCTR is a means to monopolise it.
- Enforcing the MCTR would be dif ficult since it is neither a Law , a Treaty nor an Agreement.
- Lastly while it restricts non American arms sales in the world it seeks to promote those of America. These double standards are difficult to accept.

Indian Missile Program

99. India is facing a serious disadvantage in missile technology and is still largely dependent for tactical missiles on foreign countries. It has however developed strategic missiles.The main missile based threat is from China. If China enhances its missile capability it will raise Indian concerns as it will place it in an adverse position.The Chinese in turn are concerned with their disadvantage vis a vis USA. Further Pakistan-Chinese alliance to contain India does not rule out that the missile & related technology will not be passed by the Chinese to Pakistan.

100. In the existing system most major Indian airbases are protected by up to one full squadron of S-125 Pechora-1b SAMs. Some of the bases closer to the IB and LoC have additional deployments of OSA-AKM mobile SAMs and nearly all bases have close in defences centred around radar directed 40mm Bofors L-40/70 AAA guns and Igla-1M MANPADS. Non-military targets, such as cities and critical industrial sites are also well protected with Pechoras and Bofors AAA guns. In the past, India deployed as many as 30 V-750 Dvina launchers in 5 squadrons to defend Delhi alone. Now the Dvinas are replaced by Pechoras, but the basic concept remains unchanged.The Indian army operates several regiments of Kvadrat, Strella 10M and OSA-AKM mobile SAMs but these are intended for the defence of the army's field formations rather than for the defence of targets within India. India will continue to face a threat from air-attacks, the principal menace comes from short, medium and intermediate range ballistic missiles launched by Pakistan and China. In time, these will be augmented by stealthy cruise missiles and represent a clear and present danger to the entire landmass of India - no target, no matter how far from the borders - is immune from attack. (Dr Sanjay Badri Maharaj)

101. Indo-Chinese relations are not exactly warm after the 1962 Chinese incursion into India and occupying large chunks of its territory. The border dispute has still not been settled, and China has an eye on the state of Arunachal. It also feel threatened by the Indo-US nuclear partnership and growing military ties. Further both countries have been rivals in Asia and will continue to be so. Therefore India will seek parity with China if not a superior edge. Depending upon India's response Pakistan is likely to follow suit to gain parity with India. All three countries have a missile program that is not transparent, they do not have ballistic missile early warning (BMEW) systems and all three countries have an independent nuclear and missile programs not subject to any treaty or agreements. China is also assisting Pakistan in developing missile and nuclear technology to contain India and as a lever against USA. All this makes the area very sensitive and a risk to peace. India has entered into a nuclear agreement with USA, and relations are coming closer.

Concerns, due to Chinese growing economic and military threat has caused India to consider developing a NMD. It is considering theAmerican PAC-3 system. Also on the anvil are the Russian S-300 and Israeli 'Arrow system' (Peter Dyvad). India is also co-manufacturing the Brahmos supersonic missile with Russia.

102. Replying to the 14th report by the standing committee on defence in the Parliament the representative of the DRDO stated During the Study Visit to Hyderabad when asked bout the status of missiles being developed by DRDO, the Bharat Dynamics Limited replied as under :—

> "Integrated Guided Missile Development Programme' to design, development leading to limited series production of Prithvi, Akash, Trishul, Nag and a technology demonstrator Agni was undertaken in July 1983. Nag, Akash and Dhanush, naval version of Prithvi missile system are part of the Integrated Guided Missile Development Program. The total investment under 'Integrated Guided Missile Development Program' is of the order of Rs. 1341.2 crores. Agni-I & II and Prithvi 150 Km range, 250 km range and naval version Dhanush have already been inducted into the Indian Armed Forces. However if a comparison is made with international developments in this area India has barely begun and a lot needs to be done. Self reliance is the only answer.

103. Subsequent progress has been the forming of a Joint venture comp any in India between India and Russia to develop the supersonic Brahmos missile system using state of art technology. India has announced plans to develop a two-tier ballistic missile defence system to deal with incoming ballistic missiles. The system is to use satellites for communications and a unique two layered defensive line using surface-to-air missile for any incoming ballistic missile attack (PTI 9 Feb 2003). On 6 Mar 2009 India successfully carried out a test interception of an unfriendly missile from wheeler island off Orissa at an altitude of 75 Km. a two stage 'exo-atmospheric' hypersonic missile was used this will be in Phase-I to be deployed by 2012.This demonstrated India's technology (Times of India 7 Mar 2009). It has also been reported in the same periodical dated 10 Mar 2009 that the DRDO has begun work on a more advanced 5000 km range interceptor missile in Phase – II to be deployed by 2014. This will have to have a sp ace based surveillance system to detect a hostile missile launch which would be an advance over a long range tracking Radar (LRTR). The added advantage of this program is that it can be upgraded to develop an anti-satellite capability as a deterrent in order to safeguard national assets in space.

While USA is keen that India joins it in developing a globally ef fective missile defence it would be in India's interest to develop an independent program where use of assets is not hindered by international pacts.

Assessment

104. An overview of the global scene on missile technology is as follows:-

- The line dividing missile technology between ballistic missile and Space launcher is thin.
- Technology is dominated by USA, Russia, UK and France. Japan is developing with assistance from USA using the Delta rocket series. Second tier countries like China, Korea, Brazil and Israel have acquired technology from abroad and have further developed their own base.
- The second tier countries have a commercial interest in offering Space products to other countries as a way to earn revenue and to subsidise their own development work.
- Countries like China, north Korea, Pakistan also have a military interest, the thinking is that if technology is available to large number of countries the strategic advantage enjoyed by the north countries would be diminished. Country like China would find it coinciding with its global aim of power Pakistan would find it as a way to get even with India or punish America.
- Missile technology cannot be restricted like nuclear technology for two reasons. First the MCTR is weak and second, the need for the use of Space in the modern world is inescapable even for smaller countries. This need can be met by either legitimate means from the countries having it or by subterfuge. Both routes are acceptable in the national interests.

References

- Missile defence international regional and national implications –Routledge 2005.
- Thomas G Mahnken & Janne Nolan in ‘space powers Interests”.
- SIPRI Yearbook 2008.
- The National missile defense act dated 21 Jan 1999.
- Wilhelm Agrell in his essay (“Pre-empt, Balance or Intercept” – Routledge).
- China brief – Jamestown foundation 14 Apr 2008).
- Internet www.rttnews.com/ArticleViews.aspx?ID=895193&SMap=1.

- While USA is keen to enter into a missile defence collaboration with India it would be in Indian interest to develop it s own system in order to retain freedom and independence to deploy and exploit the technology. It must avoid the earlier pitfall of importing technology on conditional terms which would kill indigenous development.

Electronic / Cyber Warfare

105. Electronic Warfare is the use of electromagnetic spectrum to ef fectively deny the use of this medium by an adversary, while optimizing its use by friendly forces. Electronic warfare is a reality. It is required for attack as well as protection of the systems operating in electromagnetic spectrum .Traditionally, EW has three main components

- Electronic support
- Electronic attack
- Electronic protection

106. Electronic support is the passive function of detection of electromagnetic radiation frequencies to provide the information regarding detection and bearing of intercepted frequencies for threat or t arget location. This is an early warning import ant input to the operations commander .Electronic support capability is required to conduct any electronic attack.

107. Electronic attack is the active or passive use of the Electro Magnetic (EM) spectrum to deny its use by the adversary Active electronic attack includes the activities like jamming mobile & digit al radio communication systems including UA Vs, deception against surveillance, active cancellation of Fire control Radars, and EMP use. Active electronic attacks can be detected by an adversary due to the active transmission of the EM radiation in the environment. The passive use includes the use of balloons, radar reflector, Faraday cages, winged decoys and stealth technology. Stealth technology has spread into many disciplines.

108. Electronic protection (also known as Electronic Counter- Counter measure) include all activities which make enemy Electronic Attack activities less successful by means of protecting friendly personnel, facilities and equipments. Electronic protection may be active or passive. Active Electronic protection includes technical modification to radio equipment through frequency hopping etc. Passive protection includes strict discipline of operators and also by modified battlefield tactics.

109. Electronics warfare technologies are guarded very closely and therefore difficult to acquire. The technologies in the area of electronic war are developing at a very fast pace and most of them are guarded / kept secret.Therefore preparation has to be done against new threats and this is a dynamic process. Some of theTechnologies relevant for EW are:-

- Technology for detection of real-time Radio Frequency threats and their identification and location
- Multiprocessors for advance missile approach warning
- Multiband detectors to enable programmable EW receivers
- Decoy base countermeasures in the entire EM spectrum
- Counter measures against surveillance acquisition base on LASERS
- Software for code breaking on real-time basis the command and control signals
- Efficient HF/V/UHF antenna with super conductive arrays for jamming
- Electromagnetic bombs for suppression of enemy air defence
- Technology for recognition and tracking of signals
- Miniaturization of EW equipment with compact power source for deployment in the vehicle based systems
- Compact Systems suitable for operation in bunkers with remote control
- Multi-function, multi spectra warners
- High speed ,wideband receivers based on Gallium Arsenide technology
- Modeling and measuring RF signature to reduce platform vulnerability
- Stealth technology for various equipments

110. Future wars would be starting with information warfare to gain advantage because most of the milit ary as well as economic activities are heavily dependent on computer networks. India has large number of professionals and comp anies engaged in software development. R&D has to be done for the protection of systems from Cyber att acks. In order to have asymmetric advant age there is a need to develop Cyber offensive and Defensive capability. Considering the talent and experience available in the country India is in a position to take world leadership in this area.

111. Cyber war has become a reality .Pent agon logged more than79, 000 attempt s for intrusion into their computer system in 2005 and 1300 were successfulThe Chinese hackers penetrated US State Department computers in several parts of the world. Some of the well known Cyber attacks in recent times are given below:-

(a) There was cyber attack on Estonia by Russian hackers. The attackers brought down the Government website, banks and telephone networks

(b) Use of cyber war in conflict between Russia and Georgia

(c) Indian site of the Ministry of External Affairs was also hacked recently.

Though many books have been written on this subject but there is very little authentic dat available on this in official records from different countries. Inmost of the countries it is a matter of concern. India has large number of highly skilled software experts and this is a niche area for future development to protect our systems.

Recommendations

112. It is recommended that technology must be developed/ enhanced for the following:-

- Composite material & metal research and Nano technology.
- Laser & guidance (GPS) for weapons.
- Image intensifiers and Image illuminators.
- Radar technology for long range , synthetic aperture Radar.
- Hermetic sealing and ventilation for NBC capable vehicles.
- Develop hybrid engines and electro-drive technology upto 1000 HP.
- Emission and acoustics control to reduce thermal signature.
- On board Enhanced control Systems-control.
- Robotic engineering for unmanned combat vehicle.
- Secure global positioning systems.
- Electronic miniaturisation.
- Miniaturisation of radio and video technology.
- High quality portable batteries or alternatives with re-chargers.

- Small arms technology with inbuilt laser range finder.
- Ammunition technology.
- Portable GPS and target designator for bringing down heavy fire.
- Research in biotechnology to enhance resist ance to disease, reduce pain and increase stamina.
- Advanced technology in textile to develop water repellant, breathing, fire resistant and bio resistant type of textiles.
- All type of Sensor technology.
- Sophisticated Sonar for anti – submarine defence.
- Material technologies for hull to make it lighter.
- Material technologies for "stealth" (paints etc).
- Material technologies for electronics for Chips, Radars and Sensors.
- Opto electronics for sensors, thermal imaging etc.
- Nanotechnology for possible applications.
- Development of Missile technology.
- Development of defensive systems in satellites.

Chapter 7

DIB-2025 and Emerging Technologies

1. What should be the shape of DIB by 2025 is not easy to answer. The issue was discussed and debated with persons from variety of background. Views were obtained from the defence experts of the Army, Navy and Air Force, Strategy management experts from HQ-IDS, industry and R &D persons. The future plans of the services in open domain were examined. Some of the recommendations have also come from the reports of Kelkar Arun singh and Ramarao committees and the Parliamentary committee. Forecasting done by USA through various study groups, China and European nations were also examined to get an in-depth understanding of emerging trends in technology developments.

2. The Unanimous view that has emerged is that the wars will be fought in six dimensions, staring from the high technology area and progressing down to the conventional conflict. Therefore the type of equipment and systems would also undergo major change. Therefore in addition to immediate requirement of war equipment, the focus has to be on upgrading the DIB with technology.

3. The DIB -2025 should have the following characteristics:-

   a. The self reliance should improve to reduce dependence on import from 70 % at present to 30 % as illustrated below :-

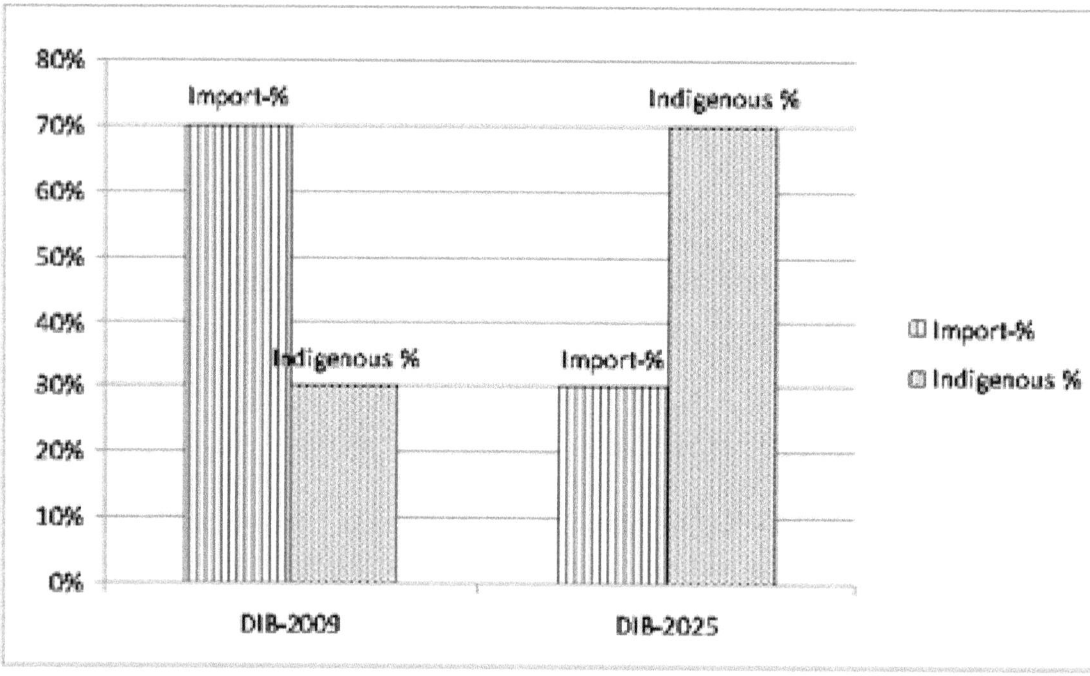

b. The private sector should develop significantly to change the structure. It has the potential to make competitive defence products. It should be possible to manufacture 25 % of the defence products in the private sector. The revised industry structure should have about 25 % of the total indigenous production as illustrated below :-

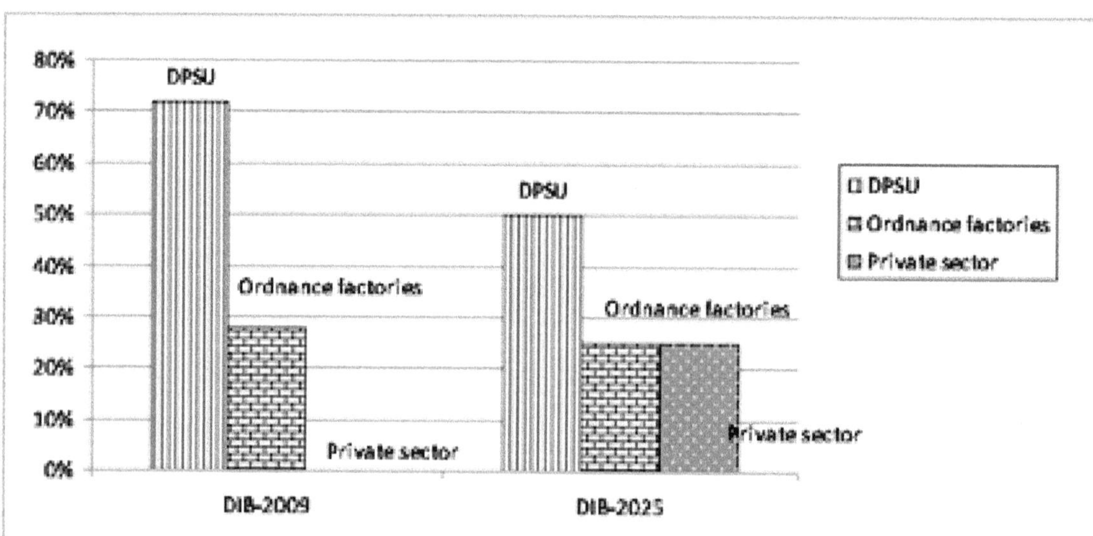

c. The DIB-2025 should have at least 10 % of indigenous products in the innovative stage and the production with laggard technology should be removed . The DIB-2025 should have the technology profile as illustrated below :-

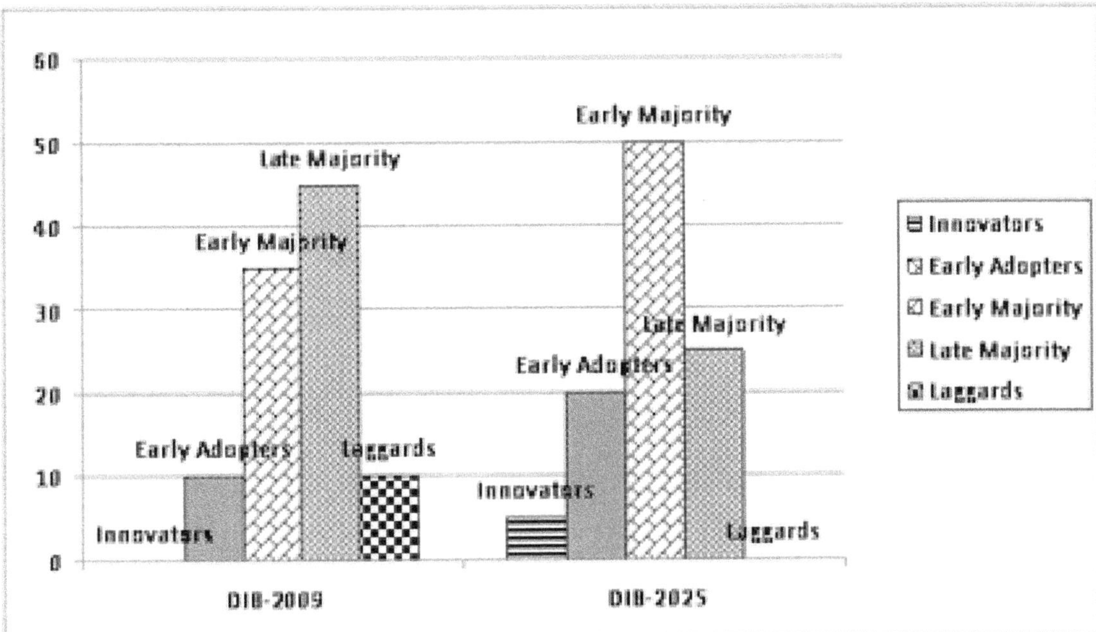

d. The R&D needs a major restructuring. At present it is dominated by the Government agencies .The S&T agencies, joint R&D, private sector and the academia in the country have the capability but they have not played any significant role in developing the cutting edge technology for the defence. There is strong need to change the existing structure to give meaningful role to other agencies. The S&T agencies in the country could contribute 20 % , followed by joint R&D (10 %) , Academia (10 %), and private sector (10 %). The existing structure for defence R&D and the desired structure for DIB-2025 is illustrated below.

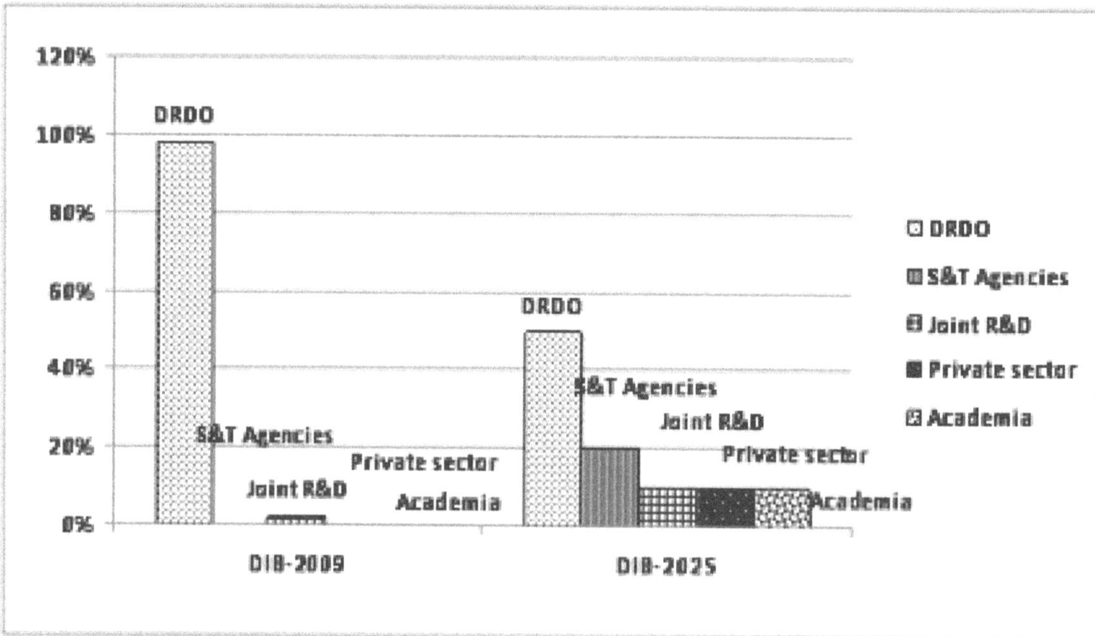

4. To make a guess estimate of the future battlefield scenario would be a risky business and something clouded in the fog of uncertainty. However if one were to see the course of events and technology as it is emerging and were to project them into the future some facets that will influence warfare can be estimated. How they will be exploited cannot be forecast. In fact wars may not be even fought the way we comprehend it now, and totally asymmetric means may be used to persuade the will of the enemy . Following things are fairly certain in future battlefield scenario.

* **Technology** - This will be in the forefront and its use would be predominant in any system. It may be purely military or dual use technology All military equipment will use technology to gain an edge over the adversary in all domain of war . High precision weapons focusing on high value targets with minimum collateral damage will be increasingly deployed. Missiles would be the weapons of first choice and consequently area missile defence (MD) would gain importance to survive a first strike.

* **Information** - Use of information and its exploitation will play a vital role. The time frame between sensors to shooter will have to be short both at the tactical and strategic levels. Therefore information superiority based on secure networks will be of importance. Electronic and Cyber warfare will precede any conflict. This will involve use of miniaturization in vision and intelligence gathering devices.

* **Infantry** - The infantry soldier will be a far cry from that of today. He will be better equipped with capability of night vision, communication from tactical to strategic level in terms of audio and video communication. He will be in a position to direct quantum of firepower commensurate to the situation in real time. His physical prowess may also be supplemented along with ability to withstand stress and immunity to alignments. His survivability would be extremely high.

* **Asymmetric warfare** - there will be increasing political groups of power in a multi ethnic society expressing their will and using force to secure their rights by unconventional means often by non state actors. In all probability the concept of western warfare of neutralizing the enemy's power centers may give way to asymmetric warfare where the Achilles heel of the enemy is attacked. The shape of the battle would be unpredictable and may be in the domain of disabling the communication systems, essential services, disrupting commerce or attack on high value targets in depth by sabotage. Asymmetric warfare in urban areas in all likelihood would be used as combat-blackmail. Use of nuclear (dirty bomb) chemical and biological warfare cannot be ruled out and would entail the armed forces to have a capability of disaster management, relief & rescue and refugee management. What will be form of warfare is difficult to guess.

* **Logistics** - logistics is the backbone of all military operations. In an environment of high cost technology, inventory costs will be high and Supply chains will perforce have to be established using best businesses practices to reduce costs and velocity logistics will have to be implemented. Joint operations in future will be the rule rather than exception. This would mean use of joint resources in an institutionalized manner There will be increasing participation of the private sector in defence. Out sourcing of products and services will be utilized to cut cost and enhance efficiency.

5. Military operations will be in full public view with all legalities transparent to the world. Military and political leaders will have to justify their actions and this will make their task more difficult. Wars will not be limited to the parties opposing each other but would involve other countries in a highly interdependent world networked together with commerce, communications, natural resources and energy. Therefore world opinion will have a tremendous influence on the warring parties and greatly limit their options.

6. To meet all situations would entail perfecting all equipment in all domains, a Utopian idea and rarely achieved. Our projections into the future should be based upon :-

- Basic research - to be carried out in premier institutions including cooperation with other countries. Research centers and academic institutions will have to be co-opted as a defence budgeted function.
- Equipment development and regular upgrading the same using technology . All technologies may not be available in the country and therefore some portions will have to be imported. Dual use technology will have advantage of economy of scale in production and earnings from selective exports.
- Achieving leadership in niche areas of our strength. It would mean research in high cost areas with a low percentage of success. This would be the biggest strength and would afford superiority in battle, commerce and international spheres. In the Indian context it would mean exploiting Info-technology and Space based programs.
- Adopting an integrated approach using all national resources of defence, private sector, academia, S&T and other agencies in a coordinated manner.
- Analysis of asymmetric threats and ways to neuter the same.This means developing versatile forces that are highly mobile to contain the damage and also developing asymmetric strength against the adversaries to dissuade or persuade them.

7. Barring the last two these fall in the domain of DIB. As a matter of policy the national security objectives must be spelt in order that policy to implement the above can be formulated. National capability desired in the future should be able to deter the adversary from any adventurism. Actual conflict in military terms may or may not follow and usually will be the last resort considering the tremendous fallout and international pressures.

8. The study has taken into account the inputs from the following to assess the desired DIB by 2025 :-

- User's perception of future requirements
- Strategy of experts from HQ-IDS
- Technology perception of DRDO
- International technology trend and forecasting
- Decisive factor in future wars.
- Future wars may be fought in six dimensions and therefore strategy , systems & equipments would undergo major change.This would require Network Centric War capability.

- Common thought has emerged that future DIB would be technology driven and Technology superiority.

9. Keeping the above in view the important observations are:-

- Future Combat forces should be :-
    - Lighter
    - More mobile
    - More lethal and precise
    - Aware of the battle space in real time through communication system and network centric war.
- Future platforms should be modular to enable up gradation of weapons and controls.
- Future DIB should have the capability to absorb the innovative imported technology and take it forward to become world leaders. It is feasible, Japan has already done it; and China is doing it now.

10. The description of required technologies for the future is of a generic nature, since specific cannot be identified. S pecific technologies have been also identified where developments have already taken place. There may be overlap in some of the technologies which is inevitable in technologies at the conceptual stage. The list would require modification to take into account the technological breakthroughs in future. The list would also need further classification to group-similar technologies together for doing Research and Development by specific institutions in the country . For building DIB some of the critical technologies which have been identified are as follows.

(a) Cyber warfare, protection and attack.

(b) Electronics war, electromagnetic attack, protection, hardening of systems.

(c) Space war, remote control, GPS, imagery.

(d) Major war equipment system integration, LCA, Tanks, warships , submarines, Missiles.

(e) The surveillance sensors.

- Radars
- SAR

- Electro optical systems
- UAV, mini and micro
- Aerostats

(f) Precision weapons

- Rockets
- Missiles
- Seeker heads/ homing devices
- Torpedoes
- Remote Guidance and fire control of our system and the enemy system
- Smart ammunition

(g) Communication

- Networks
- Data links
- Security systems / firewalls/ frequency hopping
- Cryptology
- Jamming systems

(h) Information technology

- Info warfare, disabling commercial information /control networks
- Embedded softwares

(i) Basic electronic components, microprocessors, VLSI , semiconductor foundry

(j) Directed energy weapons based on

- Lasers
- Electromagnetic spectrum
- Other form of unconventional energy

(k) Target identification in real time dynamic framework in all dimensions

- Enemy
- Friend
- Neutral objects

(l) space based capabilities for

- Remote sensing ,surveillance
- Communication
- Navigation
- Weapon delivery , control & interception

(m) MEM s technology to reduce the payloads on board UAV, Aerostat &other platforms

(n) Protection technologies against different types of attacks

- Nuclear
- Biological
- Chemical
- Others

(p) Stealth technologies based on

- Coatings
- Materials
- Design to reduce electromagnetic signature
- Deception / Optical illusion
- Electromagnetic systems

(q) Miniaturization through Nano-technology in

- Items carried by soldiers
- Weapons
- Communication systems

- UAVs
- Sensors
- Energy sources / batteries

(r) Night vision technologies

(s) Engines / energy source for mobility

- On ground
- Water / underwater
- Air / space

(t) Material

- High performance alloys
- Ceramics
- Biotechnology
- Nano materials
- Composites

(u) Miscellaneous

- Network centric war which has components spread over communication, IT and sensor technologies.
- Simulators for accelerated testing of armaments and ammunition.
- Air to air refueling
- Equipment and items for high altitude warfare
- Urban warfare against terrorist

11. For asymmetric advantage, the areas where India could be world leader by 2025 are:-

(a) Cyber warfare,

(b) Electronic warfare

(c) Communication and network centric warfare

(d) Information Technology

Technology Absorption

12. Technology is one of the most important components for building the DIB.The options are to use the technology developed indigenously or to import the technology .At the present moment our DIB lags behind technology in many areas. This is reflected by the need to import large number of defence equipment. Technology development is a slow process requiring efforts in the right direction. Technology transferred is invariably out dated and done at a time when superior technology is already on the anvil. Even when we import technology, it is not absorbed automatically by the recipient. Lot of deliberate effort is required to absorb the technology. If the technology is not absorbed fully the dependence on import becomes perpetual. Indian production systems in the private sector, public sector and the defence sectors have not been able to absorb the technology fully in most of the transactions. There are different stages at which technology may be absorbed as given below:-

- Ability to assemble the imported major assemblies in SKD (Semi Knocked Down), but no indigenous manufacture
- Ability to assemble from component stage i.e. CKD (completely knocked Down components)
- Ability to manufacture some of the components indigenously
- Ability to manufacture most of the components indigenously
- Ability to manufacture all the components indigenously with local materials
- Ability to design the product to suit the indigenous equipment and materials
- Ability to redesign the product to meet the customer's requirements
- Ability to redesign and produce the product of higher specification /version
- Ability to innovate new technology to make new product to meet the emerging needs. The innovative products / technologies make us the world leaders.

13. The highest aim of DIB should be the stage 9, where we are able to innovate new products to meet the emerging requirement in future. It may not be feasible for all the products we need, but we should definitely target on few strategic areas where India should be the World leader. The purpose of this study is to identify few areas where we could be world leaders by 2025. It is a long term strategy and the efforts have to start now. Japan

was not a country known for innovation, technology development or quality product .It was further handicapped by indigenous availability of major raw materials. In spite of all the handicaps, Japanese made a dedicated efort in absorbing the imported technology at that time. The technology for transistors was developed by the Americans in 1950s. It was imported, absorbed improved by Jap an and put to production of Transistor radios which were exported to USA. The Japanese reached the 9 th stage of technology absorption and development in less than a decade in the case of transistor technology Even at that time Japan was not known for any major technologies or products patented by them. The key lesson to learn is that *it is possible to absorb imported technology and build further on it to become a world leader.*

14. In the private sector in India technology absorption has been poor in most of the sectors and similar is the situation in the defence production. W e imported technologies repeatedly for various equipment. W e have not demonstrated the ability to upgrade the technology by our own ef forts by designing and producing products of international competitiveness in terms of performance, quality, price and delivery. The only exceptions have been the breakthroughs in the area of sp ace and nuclear power . T echnology absorption will not happen by it self unless it is backed by the commitment of the top management in the organization and appropriate funding at the budget stage.

Outsourcing of Product and Services

15. The defence services are required to play a very complex function of operations in national defence. While their core competency is in the milit ary operations they are dependent on a host of services to support these functionsThese functions are commonly utilized in the civil sector and a high level of expertise has been built up in these narrow but special fields. This has resulted in enhanced efficiency and cost reduction.The infrastructure is fast developing in the country enabling the service providers to extend their reach to remote areas. This affords an opportunity to the service providers as well as the defence services to avail of these services at optimum cost and ef ficiency. A majority of these services relate to the logistics support for obvious reasons. The areas where there could be a possible use are listed below:-

(a) R& D on technology that has possible military application in future.

(b) Product development and production process improvement to enhance its life.

(c) Research in basic science.

(d) Research on logistics and management techniques to provide support at reduced cost.

(e) Maintenance and overhaul of vehicles and equipment that have civil end use.

(f) Transport and warehousing facilities.

(g) Maintenance of military infrastructure.

(h) Provision and placement of fuel and supplies.

(i) Medical facilities in conjunction with state run / recognized hospitals in times of war to handle excess casualties.

16. Out sourcing of product and services is possible but implementation needs to be backed up with change in policy, organizational structure and manpower policy. While the savings in terms of manpower expenses (including retirement benefits) and maintenance cost of infrastructure is immense there would be intense resistant to change as it would educe the existing organizations and their internal influence. In case of the defence services the problem of manpower is not so accentuated as the same can be eventually redeployed to the fighting arms within the existing cost. Thus it amounts to organizational restructuring.

10. In order to implement outsourcing certain elements can be implemented immediately by providing a budget for the same - serial (a) to (d) above. In others where manpower and organizational changes are required a deliberate process would have to be initiated after a policy decision on the subject.

Recommendations

17. Focus of production has to shift to upgrade process, technology and products constantly. This is achieved by production expertise and R&D under one head.

18. The cycle time for negotiation and implementation is very long and in many cases the technology is superseded by new technologies in the meantime.

19. In the past production has been done based on heavy import of critical components. Concerted efforts have not been made for the absorption of imported technology. This can only be achieved by budgeting separately for technology absorption.

20. Develop a coordinating agency for organization for technology management with input from the three services, Production agencies, Academic R&D , DST and DRDO. The suitable location would be in in the Mod coordinated by HQ IDS.

21. R&D at present in the government production units is ad-hoc or optional and must be integral to their role.

22. Policies frame work for DIB, Industrial licensing and Defence procurement policies have been modified recently and their imp act on DIB is yet to be seen. Policies should create an enabling mechanism and not barriers to growth in the desired direction. Bureaucratic control has to be loosened to *become a facilitator from a controller.*

23. Government defence production units must be made to function like a corporation. The lessons drawn worldwide are a pointer to this.

24. Outsourcing of Product s and services must be accepted as a matter of policy to achieve efficiency, cost reduction and save on defence establishment and pension bills.

25. Specific military technology for development have been listed in Chapter 6.

Chapter 8

Recommendations for Bridging the Gap

1. The recommendations for building the Industrial base for 2025 are summarized in this chapter. They have been grouped under the following heads and enumerated in detail in the succeeding paragraphs.

- Policy framework
- Technology Management
- Production
- Human Resource Development
- Research and development

Policy Framework

2. Policy is the bedrock of all organisational activity. It sets the aim clearly and spells out the methodology to be followed to implement the organisational aim in a uniform manner without waste of resources. In the present context policy has been the reason for the success or failure of the DIB in India. Recommendations based on the study carried out. Are as follows:-

- Reform of the bureaucratic controls is the biggest hurdle to develop DIB in India – India can come to the level of top six countries in the world if these are eased. This is a major hurdle to progress. This is highlighted by the recommendations of numerous high powered committees and Government organisation that have yet to be implemented.
- Policies should create an enabling mechanism and not barriers to growth in the desired direction. Bureaucratic control has to be loosened to *become a facilitator from a controller.*

- India needs to reduce its Import dependence of milit ary hardware from 70% to a minimum – Policy has to be announced to achieve this aim as a national objective.
- Self reliance in DIB should be a priority area since Technology denial regime will continue to manifest itself in one form or another.
- Policy on national security has to be articulated in the overall interest of the country and to imp art a clear sense of direction for all. - China has made a political commitment of a 'Grand strategy ' to modernise the defence forces, a political & National commitment needs to be made in India by way of enunciating a National security policy and policy on DIB growth.
- Establish a coordinating agency at the apex between industry,defence and ministries on the lines of BIS on priority.
- Planning based on national security policy has to be coordinated between the Planning commission, Finance ministry and the Defence ministry & Ministry of external affairs followed by the defence policy.
- The LTIPP of the services must be integrated in the planning of the Finance ministry and Planning commission.
- Private Industry must play a greater role in defence industry and the attitude must change from *'granting permission'* to *'encouraging growth'* .
- Encourage private sector production by providing level playing field and through suitable mechanisms like development order for Make category of items.
- The Research agencies in the country may be brought into the offset loop to develop dual use technology.
- The recommendations of the Group of ministers and the Arun singh committee must be implemented.
- As an interim measure the five year defence plans must be approved along with the national Budget of the first year of the plan and revised each year thereafter.
- The cat alyst to achieve the above can only be a CDS who must be appointed without further delay.

Technology Management

3. Reduce technology development cycle - After a National policy the process of planning and positioning a sound Technology management Plan is important. More than developing a DIB the management of the same takes precedence.This has been a major grey area in• India leading to delays in negotiations and consequent delay in implementation. By the time technology fructifies it is already obsolete, defeating its very purpose.

4. Enhance technology absorption - Imported technology absorption has been very limited .In most of the cases the production has been done based on heavy import of critical components. Concerted efforts have not been made for the absorption of imported technology. There is no budgeting for technology absorption in the production units. There are hardly any cases where production units have absorbed and upgraded the technology to design and manufacture higher version of the products. In most of the cases the design & R &D component was missing in the imported technology package.

5. Recommendations for technology management are given below:-

- Technologies which are dif ficult to obtain should be leveraged through multi-channels on a case to case basis.
- Appoint a central coordinating agency for defence technology management - Inputs from the three services, Production agencies, Academic R&D , DST and DRDO. The suitable location would be in IDS representing the end user and all stake holders. It could be organised on the lines of BIS.
- Organisational reforms need to be implemented to absorb imported technology.
- Investment in technology development has to be in the first quadrant of the technology development cycle discussed in Chapter 1 ie Innovation phase - This will impart advant ages of leadership in technology and bargaining power . Commercial spin off benefits are inbuilt in success in any area falling in this quadrant.
- Emerging requirements of Operational capability and technology are known to defence HQ's and factored in the L TIPP these must be implemented with Policy and Budgetary support.
- DARPA has been very successful in developing defence technology and there is a need to examine the feasibility of adopting a similar model with modifications in India.

- Converting the DGOF units into corporations be carried out - The benefits of this has been demonstrated worldwide even in hard-line socialistic states like Russia and China. This could form a part of national dis-investing policy of the PSU's.
- The focus of technology must shift from conventional hardware to informalization in order to orient all defence technology to Digital and Network based.
- The potential procurement plan of India must be leveraged to obtain technology - Considering India's huge defence budget it has a potential to leverage bidders to meet its requirements.
- A study may be carried out to examine how India can exploit the economic downturn currently being experienced in developed economies to obtain high technology.
- There is a very strong case of establishing a functional DOFA – it must be an independent office with a separate budget. It must perform two roles ie as a facilitator and as a Monitoring agency. Both roles complement one another and are best handled by DOFA.
- The policy on Offsets should be reviewed regularly based on experience of its application.
- Transparency is of great value in making utilization of offsets a success and entails giving complete project information, selection criteria, future expansion, time frame and other liabilities in the RFP stage.
- The time frame for according an approval or Rejection must be fixed and is not open ended.
- The investment level of 26% must be maintained till experience and mutual confidence is established - This must be viewed in the light of emerging realities of global recession, increasing mutual interdependent world, India's promise of further business and the fact that foreign investors need India as India needs them (the world is growing flat). The foreign investor need have no fears if they intend to honour their commitment in the long term. The investment level could be raised up to 49% in case where the supplier intends to establish manufacturing facilities from scratch. A conditional clause could be added in the policy that on successful completion of their offset commitments an investor may apply for enhancement of their share with evidence of future business potential. This would be assessed on case to case basis. The government must make it mandatory to supply the reasons for approval or rejection.

- Offset trading/ transfer could be made transferable subject to the transferee using them for defence industry only.
- The list of areas where of fsets are required should be evolved jointly by users, CIDS, DRDO and other Research agencies. It should be revised regularly to make it more specific and to ensure that an investor 's proposal is not shot down at the eleventh hour.
- The need to apply multipliers in giving credit may be examined to give direction for flow of offsets in the desired direction.
- Capital acquisition must be known from the st age that approval is granted for it. (Most of the secret information is available worldwide in trade journals and research papers, and is classified as secret for home consumption).
- The defence Minister must have the final say on planning and executing the defence budget once allotted by the Ministry of finance.
- The budgeting and procurement procedure, based on the works procedure, needs to be reviewed.
- The CAG must carry out an *annual review* of the defence spending and report the same to parliament.
- All defence procurement of capital equipment must be taken up as a project under an appointed leader – Supported by a team who will be accountable from induction to total deployment. Further allocation of funds or withholding the same must be conditional on performance.
- The procurement system based on L1 vendor needs to be reviewed.
- The DGOF have been outsourcing supplies since inception but this has not resulted in building a second tier of suppliers & manufacturers due to Industrial policy until 1991. Policy Decisions therefore must be taken holistic manner.
- An annual review on progress of p articipation by private sector and reasons for delay must be included in the annual report of the Ministry of defence.
- Effect of variations in rate of exchange of foreign currency must be spelt out in the DPP for the benefit of Indian private sector..
- DOFA in its proposed 'avtar' must deal with issues of offset banking.

Production

6. Production is at the heart of building a DIB. It needs a coordinated ef fort at the national level between all resources for success. Most western countries post WW 1 1 made a conscious to develop this capability and protect the technology developed by them to reach where they are today. The major recommendations are summarised below:-

- Restructure the ordnance factories to make them competitive through the following:-
  - Greater autonomy by conversion to corporations.
  - Carry out BPR (Business Process Engineering) to rationalize their product profile.
  - Restructuring may be done through setting up holding companies having production, design and R&D under one CEO
  - Appropriate % of the capacity may be reserved for production of strategic products .
  - Introduce the concept of competition wherever possible among the units in government sector also.
  - Make the organizations lean for high performance. Their turnover to manpower ratio is very poor when compared to any other leading manufacturers in the SIPRI top 100 companies.
- Change the mandate of Ordnance Factories to include
  - Product development through in-house R&D (surprisingly it is almost absent in the existing system)
  - Technology absorption with clear understanding that mark-2 of the product has to be developed in-house.The process of technology absorption has to start from day-1 and should be backed by suitable funding and manpower.
  - Ordnance factories should spend 5 % of their turnover on R &D.
  - Enhance R&D in the PSUs to a level of 5% of their turnover for product development as well as technology absorption. They should also have mechanism for sponsoring projects on other institution in public and private sector including the academics.

- Encourage private sector production by providing level playing field and through suitable mechanisms like development order for Make category of items.
- Focus of production has to shift to constantly upgrade technology.
- The cycle time for negotiation and implement ation is very long and in many cases the technology is superseded by new technologies in the meantime. A study be carried out to examine how to cut down the negotiation-approval cycle time.
- Large government comp anies contracting Of fsets should be encouraged to farm out small portions to the SME in order to strengthen the DIB.
- SME in India hold great potential for forward integration and must be encouraged with long term orders.

Human Resource

7. The Best of resources and technologies are rendered null without the requisite human resources to exploit them. While planning for a modern DIB concurrent planning for requisite technically qualified manpower from educational institutions and to retain talent has to be done. Recommendations on Human resource are given below:-

- Convert Brain-drain to Brain-gain - Brain drain is a serious problem in high technology areas and requires a deliberate policy to prevent it. While the private sector is not encumbered with dep artmental rules in government sector it needs a rethink. Suitable measures may be taken for retention of professional expert s in defence R&D.
    - For high tech projects develop mechanism outside the existing rules to induct highly qualified/ experienced persons on project basis for a limited time.(Sam Pitroda did it successfully in C-DOT)
    - Expedite the operation of National Defence University (NDU) - for education and training of manpower in all the areas including strategy, technology management, production and R&D.
    - MOD should develop policy for compulsory upgrading the educational and technical skills of key manpower engaged in production or R&D (At present it ad-hoc or optional)
    - Delink terms of service of production units and follow the corporate principles.

- Need to educate the Bureaucracy to act as facilitator rather than as a dispenser . Banks and Postal services and railways have implemented this successfully.
- The DRDO seeks ideal work conditions and facilities it does not promise performance in return. Therefore incremental benefits must be conditional on performance.
- Study leave of any kind must be based on demonstrated cap ability in the past so as to be a further build up on the same. Also leave has to be concurrent with a commitment to benefit the organization and therefore need for a bond to serve four years further in respect of the scientific community. .
- Work and living conditions must be at par with the best and in consonance with the income of the government as part of hygiene factor and also a national commitment of sharing the benefits with all strata of the country.
- Internal communication within the campus must be made available to maximum possible personnel.

Research & Development

8. The R&D lead in India is taken by the DRDO for defence related technology. The Defence R&D should be more focused as recommended by the committees headed by Ramarao, Kalelkar and the recommendations of S tanding committee of Parliament 14 th report. DRDO should focus on thrust areas instead of large number of area to get better results. The criteria could be:-

- The areas where technology is not available and it is strategic (lists identified in chapter 2)
- Where private sector involvement may not be desirable from national security point of view.
- Target few technology areas where India should be world leader in time to come. This would consolidate India's position to negotiate technology transfer in future as an equal partner. The chosen areas could be the one where we have strength and the probability of success is high.

9. Majority of the development work may be done on project / mission mode. The projects may be planned in consult ation with the team for more realistic cost and time frame which could be used as a base for monitoring the progress internally by the team and externally by the sponsors of project.

10. Introduce the R&D teams to the concepts of Project Management and implement the use of modern Project Management techniques so that there is better / realistic planning, implementation and control of R&D Projects. While this exists on paper the implementation is weak. Projects are characterized by definite start & end date with inherent constraints of resources like funds and manpower . Many R&D project s have been run as operations which continue indefinitely.

11. R&D is a high risk area .All the projects are not likely to succeed in R&D; therefore there should be clear-cut policy guidelines for dropping a project instead of carrying it on indefinitely. Witch hunting must be avoided; otherwise high risk projects would not get the desired encouragement.

12. The projects which fail should be also documented for the learning process and this should not be used for witch hunting.

13. DRDO should focus more on technology development and system integration.

14. Introduce the concept of concurrent engineering where the production, design team and users work simultaneously and not in a sequential order.

15. The technology transfer mechanism from DRDO to production units should include a more comprehensive technology package (product design, process technologies, choice of production technology/ equipment and materials & quality assurance etc. depending on the technical competence of the recipient of technology).This would cut down the time for the product development to reach the final users.

16. Identify more projects for joint development with suitable Indian organizations and international organizations abroad.

17. Develop institution like DARPA within MOD / HQ-IDS for sponsoring cutting edge technology development projects in India on appropriate

- PSU
- Private sector
- CSIR labs
- Other departments
- Academic institutions of repute

18. Develop strong and long term linkages with Academia to:-

- Increase the S &T Pool in the country
- Attract talented persons in R&D
- Set up large number of centres of excellence with linkage to suitable labs.
- Offer large number of projects related to dual use technology at under graduate, post graduate, doctorate and post doctorate level. Existing interaction is sub-critical.

19. Change the mandate of Ordnance Factories to include:-

- Product development through in-house R&D (surprisingly it is almost absent in the existing system)
- Technology absorption with clear understanding that mark-2 of the product has to be developed in-house. The process of technology absorption has to st art from day-1 and should be backed by suitable funding and manpower.
- Ordnance factories should spend 5 % of their turnover on R &D.
- Enhance R&D in the PSUs to a level of 5% of their turnover for product development as well as technology absorption.They should also have mechanism for sponsoring projects on other institution in public and private sector including the academics.

20. It is recommended that specific technology must be developed/ enhanced for the following:-

- Composite material & metal research and Nano technology
- Laser & guidance (GPS) for weapons
- Image intensifiers and Image illuminators.
- Radar technology for long range , synthetic aperture Radar
- Hermetic sealing and ventilation for NBC capable vehicles.
- Develop hybrid engines and electro-drive technology upto 1000 HP.
- Emission and acoustics control to reduce thermal signature.
- On board Enhanced control Systems-control.
- Robotic engineering for unmanned combat vehicle.

- Secure global positioning systems.
- Electronic miniaturisation.
- Miniaturisation of radio and video technology
- High quality portable batteries or alternatives with re-chargers
- Small arms technology with inbuilt laser range finder
- Ammunition technology
- Portable GPS and target designator for bringing down heavy fire
- Research in biotechnology to enhance resist ance to disease, reduce pain and increase stamina)
- Advanced technology in textile to develop warer repellant, breathing, fire resistant and bio resistant type of textiles.
- All type of Sensor technology
- Sophisticated Sonar for anti – submarine defence
- Material technologies for hull to make it lighter
- Material technologies for "stealth" (paints etc)
- Material technologies for electronics for Chips, Radars and Sensors
- Opto electronics for sensors, thermal imaging etc.
- Nanotechnology for possible applications.
- Development of Missile technology
- Development of defensive systems in satellites
- Academic Institutions to be incorporated in the development of military technology as a Budgeted function by the Ministry of Defence.
- The recommendations of the NKC relevant to the DIB must be followed

21. Road Map for Implementation of Recommendations.

Some of the recommendations can be implemented within 2 year in phase -1. The recommendations which require further consult ations for implement ation may be done within 3 years in phase-2. Some recommendations which call for a fundament al major change may take up to 5 years for implementation in phase -3.

22. Phase -1 for implementation within 2 years

Policy & Procedures

- Bring about transparency in purchase procedures to encourage the growth of DIB.
- Declare clear cut Mission to reduce imports dependence from 70 % to 30 % by 2025.
- Strengthen DOFA, make it proactive to give direction to leverage offsets to get technologies which are not possible otherwise.
- Procurement procedure based on works procedure, needs to be revised.
- The concept of L1 to be reviewed.
- Encourage private sector by:-
- Implementing RUR policy
- Involving them in Make items at early stage
- Introduce the concept of Project management in R&D projects for better planning, execution and control (train large number of trainers for this)

Technology

- Introduce the concept of concurrent engineering to cut down the cycle time for product / technology development.
- Encourage CII/ FICCI to set up a Consortium of leading manufacturers to undertake defence related R&D projects. This would initiate defence R&D culture in the private sector.

23. Phase-2 for implementation within 3 Years

Policy and procedure related

- Change the mindset from bureaucratic control to facilitation in bureaucracy through declaration of policy.
- Encourage exports to make the production competitive.
- Policy on national security needs to be articulated.
- LTIPP must be a joint process between MoD HQ IDS, MoF and Plg commission
- The Def Min must have final say on expenditure in the MoD after allocation
- Annual review by CAG and report presented to Parliament to encourage transparency.
- Capital acquisition to be taken up as a project under a designated leader
- Recommendations of various high level committees must be implemented
- Expand / upgrade defence education & training facilities to meet emerging demands .

Technology Related

- Setup Defence S&T Agency / Commission under MOD but outside of DRDO for:-
  - Technology forecasting
  - Technology assessment
  - Technology strategy (develop, import, jt. development)
  - Promoting innovation within/ out side DRDO
  - Identifying high tech companies for jt. development / mergers
- DRDO should rationalize priority areas of R&D as recommended by Ramarao & Kalelkar committee.
- Set up an autonomous organization within MOD for encouraging innovation by sponsoring high tech futuristic projects (similar to DARPA of USA & NMITLI of CSIR)

- Plan missions for developing technology in priority areas to be executed Within DRDO and Other S&T institutions e.g. DST has a Mission on Nanotechnology
- Stress more on basic innovative technology as a policy because technology denial would continue
- Phase out products in the Laggard phase of technology as a matter of policy

Production Related

- Set up an apex body/ commission in MOD to facilitate the long term growth of DIB under Secretary Defence
- Capital acquisition must provide all details from RFP stage and level field for all suppliers. PSU and private sector (Indian) must be at par.
- Policy on manpower in prod units need to be reviewed- corporatisation is the answer
- Giving development projects to private sector with cost sharing
- Develop national data base for encouraging subcontracting of products / services. This could be located in the Ministry of Industries under DIPP.

24. **Phase -3 for implementation within 5 years**

Technology Related

- Develop mechanism in MOD for regular interaction with other S&Tagencies for funding dual use technologies
- Develop mechanism for interaction with reputed academic institutions in the country for
    - Setting up / up - gradation of "Centres of Excellence"
    - Deputing experts from defence to implement Research projects
    - Developing large number of technology projects of dual use technology at under graduate, PG and Doctorate level in academia.

Production Related

- Make Ordnance factories lean & competitive through
    - BPR to rationalize their operations
    - Restructure them into holding companies having production, design and R&D under one CEO
    - Make them lean by rationalizing manpower requirements.
    - Introduce competition where ever possible
- Modify the mandate of Ordnance factories to include in-house R&D for next generation Product development & technology absorption with adequate resources.

Monitoring Mechanism and Milestones

25. A strong mechanism is required under MOD for monitoring the growth of all the components of DIB (production, technology and policies). The mechanism should have the following characteristics

- It should be independent from production, R&D and the facilitating agencies.
- The mechanism should be able to underst and & appreciate the requirements of the user defence services.
- It should have the capacity for networking & power to take corrective measures
- The ideal location could be in HQIDS under MOD provided it is strengthened suitably to undertake this work.

It should monitor the progress of key p arameters for the growth of DIB. Which are given below:-

Parameter	Base line - 2009	Target 2015	Target 2020	Target 2025
%age of imported armaments	70	70	50	30
Private sector production share *	negligible	2 %	10 %	25 %
Setting Centre of excellence in academia @	15	50	100	200
Laggard production	10 %	Nil	Nil	Nil
Innovative phase production	0	0.5 %	2 %	5 %
World leadership in technologies	Nil	>Cyber War >Secured Net work >GPS >Electronic Warfare >NCW >Space	Advanced Technology	Advanced Technology

* The private sector needs time to assimilate the mechanics, once mastered it is a simple question of increasing the scale of production.

@ The academic research must be supported by the MoD by allocating Rs 500 Crores annually to fund or subsidize defence related projects. This will be a tremendous motivation at both ends.

26. The Mile stones for equipment related modernization are listed below The time frame required is different from that of policy since development, testing and evaluation is a long drawn process before deployment of equipment. The time frame has been taken keeping in view the development cycle. Further the medium term is basically a bridge between the short term and long term and is therefore flexible. The aim is to provide a road map for development. While these are considered milestones they are subject to modification on the merit of each case.

Recommendations for Implementation			
	Short term (less than 10 years)	**Medium term (10-15years)**	**Long term (15 years & beyond)**
1	Indigenise full Combat vehicle manufacture	Develop advanced vehicles & Robots	Robotic combat vehicles
2	Develop secure communication systems up to front line sub-unit commander.	Deploy INSAS	Enhance INSAS
3	Develop night vision	Remove night blindness	Enhance night vision quality to all Operations
4	Develop LCA, MRCA & Drones	Deploy LCA, MRCA & Drones	Advanced Aircraft with long sustenance including unmanned aircraft.
5	Develop hypersonic missile technology.	Deploy missiles	Fully integrated Missile Defence
6	Develop tactical missile	Deploy missile	Integrated 2nd Artillery
7	Warship design and manufacturing capability to be developed.	Advanced Naval technology	Blue water capability
8	Develop Radar technology to cover complete spectrum of operations.	Deploy Radar	Integrated Radar defense
9	Develop and deploy precision tracking and guidance systems.	Deploy systems on land sea & air	Integrate system on land, sea & aerospace systems
10	Initiate research in military application of information technology including embedded technology.	Enhance product performance	Integrate "Decision-maker" with "Shooter"

contd/-

Recommendations for Implementation			
	Short term (less than 10 years)	**Medium term (10-15years)**	**Long term (15 years & beyond)**
11	Develop Internet platform	Test Internet in military operations	Independent & secure military Internet 'Bharat Net'
12	Develop satellites based reconnaissance and Imaging systems.	Harden satellites	Defensive & offensive measures in Space to safe guard assets
13	Develop GPS capability for aviation	Extend GPS to operations	Independent GPS.
14	Establish cyber-war research	Secure national computer systems	Full cyber war capability
15	Implement Standardisation and establish commonality of logistics and administrative assets.	Initiate utilisation of common resources	Integrated logistics

Appendix A

Country Offset Policies

Australia

Title of Policy:	Australian Industry Involvement (AII) and Defence Materiel Organization (DMO) – not termed offset but Strategic Industry Development Activities (SIDAs)
Agency Handling	(DOD) Defense Acquisition Agency
Part of Procurement Decision	No
Offset Sector	Civilian and military
Minimum Value of Contract	A$2.5M foreign content/any tender of A$5M
Minimum Offset Required (%)	Maximized where cost effective
Term	Not defined
Multipliers	None in policy
Penalties	During project, strict SIDAs review, if not completed, not credited: must fulfill obligation
Focus	Local content (Australia & New Zealand), SIDAs
Direct vs. indirect	Both
Eligible Offset Activities	Local production, R&D, tech transfer, training, export sales,infrastructure, collaborative ventures
Website	Defence Materiel Organization: http://www.dmo.defence.gov.au/DMO function. cfm? function id=10

Belgium

Title of Policy:	Industrial Benefit in the Field of Defense Procurement ("economic compensations")
Agency Handling	Ministry of Economics Affairs
Part of Procurement Decision	Yes
Offset Sector	Civilian and military
Minimum Value of Contract	Not specified
Minimum Offset Required (%)	100%
Term	Not defined
Multipliers	None in policy
Penalties	Penalty for non-executed obligation
Focus	High technology
Direct vs. indirect	Both
Eligible Offset Activities	Co-production, direct supplies & services, tech transfer, R&D
Website	Ministry of Economic Affairs: http://mineco.fgov.be Industrial Benefit in the Field of Defense Procurement (Overview): http://mineco.fgov.be/organization_market/index_en.htm

Canada

Title of Policy:	Industrial & Regional Benefits (will not use term "offset")
Agency Handling	Industry Canada
Part of Procurement Decision	Yes
Offset Sector	Civilian and military
Minimum Value of Contract	C$2M preferred – C$100M required
Minimum Offset Required (%)	100%

Term	Not defined
Multipliers	None in policy
Penalties	Performance guarantee utilized
Focus	Economy, job creation, technology, politics
Direct vs. indirect	Both
Eligible Offset Activities	Well defined
Website	Industrial and Regional Benefits (overview): http://strategis.ic.gc.ca/SSG/ad03657e.html

Denmark

Title of Policy:	Industrial Cooperation Agreement (ICA)
Agency Handling	Danish Agency for Trade and Industry (ATI), under the Danish Ministry of Business and Industry
Part of Procurement Decision	Yes
Offset Sector	Civilian and military
Minimum Value of Contract	25 million DKK (approximately US$3.8BN)
Minimum Offset Required (%)	100%
Term	Not defined
Multipliers	None in policy
Penalties	Performance guarantee utilized
Focus	Local content (Australia & New Zealand), SIDAs
Direct vs. indirect	Both
Eligible Offset	Defense and technology similar to product purchased
Website	Policy guidelines and sample agreement: http://www.efs.dk.ati/

Egypt

Title of Policy:	No official policy
Agency Handling	Not specified
Part of Procurement Decision	Ad hoc basis
Offset Sector	Military
Minimum Value of Contract	Not specified
Minimum Offset Required (%)	Low
Term	Not defined
Multipliers	None in policy
Penalties	Not defined
Focus	Defense industry development and support
Direct vs. indirect	Direct
Eligible Offset Activities	Direct, technology transfer
Website	No Website Available

Finland

Title of Policy:	Industrial Participation
Agency Handling	Trade Ministry (& Finish Committee on IP)
Part of Procurement Decision	Yes
Offset Sector	Civilian and military
Minimum Value of Contract	Contracts exceeding 10 million Euro mandated by Parliament
Minimum Offset Required (%)	100% + marketing consulting
Term	Not defined
Multipliers	1-3 times if Finnish products are exported

Penalties	Exclusion from future bids until contract fulfilled
Focus	Participation of domestic defense industry, technology, export, internationalization of exports
Direct vs. indirect	Both
Eligible Offset Activities	Transactions that benefit economy and industries, technology transfer
Website	New Offset Guidelines: http://www.vn.fi/ktm/eng/2/2_5.htm Rules of Industrial Participation: http://www.vn.fi/ktm/eng/2/2_6.htm

Germany

Title of Policy:	Industrial Balances, no official policy
Agency Handling	The Federal Office for Defense Technology and Procurement
Part of Procurement Decision	Yes
Offset Sector	Not available
Minimum Value of Contract	Not available
Minimum Offset Required (%)	Aim is 100%
Term	Not defined
Multipliers	Not available
Penalties	Not available
Focus	German company participation from the inception of a project, balance globalization with growth of global companies
Direct vs. indirect	Both
Eligible Offset Activities	If just procurement contract, co-production required
Website	Website not available

Greece

Title of Policy:	Policy of Offsets Benefits (O/B)
Agency Handling	Hellenic Ministry of National Defence/General Armaments Directorate (GAD)
Part of Procurement Decision	Yes
Offset Sector	Military
Minimum Value of Contract	250 million drachmas
Minimum Offset Required (%)	80-120%
Term	Not defined
Multipliers	Very complex, depends on value of offset & recipient, maximum is 12
Penalties	10%
Focus	Defense industry and co-production
Direct vs. indirect	Direct
Eligible Offset Activities	Require: (co-production) local content, local subcontracting, joint venture, technology transfers
Website	Armaments General Directorate Description: http://www.mod.gr/english/index.htm (Go to White Paper of Armed Forces, Section 4.1.2.10) Offset Benefits Policy Description: http://www.3dpaper.gr/eng/defence/offsetbenefits_ def040101.shtml

Israel

Title of Policy:	Industrial Cooperation (Industrial Cooperation Benefits – ICB)
Agency Handling	Industrial Cooperation Authority (ICA), Ministry of Trade and Industry
Part of Procurement Decision	No
Offset Sector	Civilian and military

Minimum Value of Contract	US$100,000
Minimum Offset Required (%)	35%
Term	Usually 3 years, may be extended to 10 years
Multipliers	1-2 times, dependent upon type of offset
Penalties	No liquidated damages clause
Focus	Development of close, long-term working relationships
Direct vs. indirect	No distinction
Eligible Offset Activities	Subcontracts, R&D, technology transfer, investment, global market access and exposure
Website	Industrial Cooperation Authority: http://www.moit.gov.il/tamas_Mainlink_english.asp?link=bin/goto.asp?{f9966}

Korea, South

Title of Policy:	Policy of Offset Benefits (O/B)
Agency Handling	Defense Procurement Agency, Ministry of National Defense (MND)
Part of Procurement Decision	Yes
Offset Sector	Military
Minimum Value of Contract	$10M
Minimum Offset Required (%)	30%
Term	Not defined
Multipliers	0-6 times based on type of offset
Penalties	Disbarment from participating for non-adherence to offset obligation

Focus	High technology must be more than 30% of contract
Direct vs. indirect	No distinction
Eligible Offset Activities	Mainly technology transfers, also employment, equipment, purchases and other
Website	Defense Procurement Agency: http://www.dpa.go.kr/servlet/kr.go.dpa.dpacals.com.SI_COM_ Guest?DIV=12

Kuwait

Title of Policy:	Offset Program
Agency Handling	Ministry of Finance, Program Executive Office PEO
Part of Procurement Decision	Yes
Offset Sector	Civilian and military
Minimum Value of Contract	KD 1 million, sum of contracts in one year
Minimum Offset Required	30%
Term	8 years with intermediary steps of completion (50% completion in 4 years, 100% in 8 years)
Multipliers	2-10 times based on activity and sectors
Penalties	6% of total contract
Focus	Technology transfer and training
Direct vs. indirect	No distinction
Eligible Offset Activities	Expenses of joint ventures with local parties
Website	Offset Program Description: http://www.mof.gov.kw/offset/

Netherlands

Title of Policy:	Industrial Participation and Offset
Agency Handling	Ministry of Economics
Part of Procurement Decision	No
Offset Sector	Military
Minimum Value of Contract	5 million Guilders
Minimum Offset Required (%)	100%
Term	10 years
Multipliers	1-3 times
Penalties	5% of late portion, must still fulfill obligation
Focus	Technology similar to products purchased
Direct vs. indirect	Mix with original focus on direct
Eligible Offset Activities	Counter-purchase, co-production, licensed production
Website	Industrial Participation and Offset in the Netherlands: http://www.minez.nl/home.asp?page=/cmp/compensatie/brochure.htm

New Zealand

Title of Policy:	Defence Offsets Policy/Industrial Involvement
Agency Handling	Minister of Defense & Ministry of Commerce
Part of Procurement Decision	No
Offset Sector	Military
Minimum Value of	NZ$5 million

Contract	
Minimum Offset Required (%)	30%
Term	Not defined
Multipliers	1-3 times
Penalties	Liquidated damages
Focus	Stimulate growth and employment, sustainable activities
Direct vs. indirect	No distinction
Eligible Offset Activities	R&D, technology transfer, joint ventures, training, export marketing, etc.
Website	New Zealand Industrial Involvement: http://www.defence.govt.nz/scripts/public/nzii.asp#Pt3_Prog_Req_General

Norway

Title of Policy:	Offset Program
Agency Handling	Ministry of Finance, Program Executive Office PEO
Part of Procurement Decision	Yes
Offset Sector	Civilian and military
Minimum Value of Contract	KD 1 million, sum of contracts in one year
Minimum Offset Required (%)	30%
Term	8 years with intermediary steps of completion
Multipliers	2-10 times based on activity and sectors
Penalties	6% of total contract
Focus	Technology transfer and training
Direct vs. indirect	No distinction

Eligible Offset Activities	Expenses of joint ventures with local parties
Website	Guide to Norwegian Armed Forces procurement: http://balder.dep.no/fd/publ/fdhowtos/index.htm

Philippines

Title of Policy:	Countertrade (Implementing Rules & Regulations)
Agency Handling	Department of Trade and Industry through the Philippine International Trading Corporation (PITC)
Part of Procurement Decision	Yes
Offset Sector	Civilian and military
Minimum Value of Contract	US $ 1 million
Minimum Offset Required (%)	50%
Term	3 years after from execution of the contract (2 year grace period)
Multipliers	2-5 subject to the value of the desired activities
Penalties	Non-performance ranging from 5%-100%
Focus	Foreign capital equipment, machinery and services
Direct vs. indirect	Indirect
Eligible Offset Activities	Co-production, countertrade or barter
Website	Philippine International Trading Corporation: http://www.dti.gov.ph/pitc/PITC

Saudi Arabia

Title of Policy:	Saudi Arabian Economic Offset Program
Agency Handling	Economic Offset Committee (EOC) (Ministry of Defense and Aviation)
Part of Procurement Decision	Yes
Offset Sector	Civilian and military
Minimum Value of Contract	Not specified
Minimum Offset Required (%)	35%
Term	Within 10 years
Multipliers	Subject to approval of offset authority
Penalties	Best efforts but reconsidering policy
Focus	Jobs, training, technology transfer and investment
Direct vs. indirect	Mix with original focus on direct
Eligible Offset Activities	Investments in joint ventures with local parties
Website	Economic Offset Program: http://www.saudioffset.com/

South Africa

Title of Policy:	National Industrial Participation
Agency Handling	Department of Trade and Industry
Part of Procurement Decision	Yes
Offset Sector	Civilian and military

Minimum Value of Contract	Import content greater than US $10 million
Minimum Offset Required (%)	100%
Term	7 years
Multipliers	1-2 subject to type of IP
Penalties	5% (on unfulfilled performance guarantees)
Focus	Develop industry, technology transfer, job creation
Direct vs. indirect	No distinction
Eligible Offset Activities	Foreign investment, exports, R&D, technology transfer
Website	Strategic Defence Procurement Package Page: http://www.gov.za/projects/procurement/

Spain

Title of Policy:	Industrial Cooperation
Agency Handling	Ministry of Defense (Industrial Cooperation Directorate/ Management Office)
Part of Procurement Decision	Yes
Offset Sector	Military
Minimum Value of Contract	NA
Minimum Offset Required (%)	100% expected contract value
Term	Not defined
Multipliers	None in policy
Penalties	None in policy
Focus	Technology similar to product purchased, economy, domestic industry
Direct vs. indirect	Both

Eligible Offset Activities	Evaluated on case-by-case basis, prefer partnerships with domestic firms
Website	No website available

Sweden

Title of Policy:	Offset and Participation Program
Agency Handling	DOD, Defense Material Administration (FMV)
Part of Procurement Decision	Yes
Offset Sector	Civilian and military
Minimum Value of Contract	100 MSEK
Minimum Offset Required (%)	NA
Term	Not defined
Multipliers	None in policy
Penalties	Performance guarantee utilized
Focus	Strengthen domestic defense industry
Direct vs. indirect	Both
Eligible Offset Activities	Co-production, technology transfer, etc.
Website	Swedish Defence Materiel Administration: http://www.fmv.se/index.asp?L=UK

Switzerland

Title of Policy:	Defense Procurement and Offset Policy
Agency Handling	Defence Procurement Agency, DOD

Part of Procurement Decision	Yes
Offset Sector	Civilian and military
Minimum Value of Contract	50 million Swiss francs
Minimum Offset Required (%)	100%
Term	Not defined
Multipliers	None in policy
Penalties	Performance guarantee utilized
Focus	Retain domestic industry independence, overcome trade barriers
Direct vs. indirect	Both
Eligible Offset Activities	Co-production, cooperation with universities, export assistance
Website	Defence Procurement Agency (under construction): http://www.vbs.admin.ch/internet/GR/e/active.htm

Taiwan

Title of Policy:	Industrial Cooperation Program
Agency Handling	Ministry of Economic Affairs & Industrial Development Bureau (IDB), Committee for Aviation & Space Industrial Development
Part of Procurement Decision	Yes
Offset Sector	Civilian and military
Minimum Value of Contract	$50 million
Minimum Offset Required (%)	30-40%
Term	

Multipliers	1-10 times, based on type of offset
Penalties	
Focus	Upgrade industrial technology, increase quality of workforce, globalization
Direct vs. indirect	Both
Eligible Offset Activities	Local procurement, technology transfer, training, R&D, marketing
Website	No website available

Thailand

Title of Policy:	Countertrade Policy
Agency Handling	Department of Foreign Trade, Ministry of Commerce
Part of Procurement Decision	Yes
Offset Sector	Civilian and military
Minimum Value of Contract	300 million Baht
Minimum Offset Required (%)	20-50%
Term	2 months prior to end of contract
Multipliers	None in policy
Penalties	5% (on unfulfilled) performance guarantee
Focus	Enhance trade, prevent imbalance of trade
Direct vs. indirect	Indirect
Eligible Offset Activities	Counter-purchase
Website	Countertrade Policy in Thailand: http://www.dft.moc.go.th/eng/Counter/Ctpol.html

Turkey

Title of Policy:	Military Offset Policy and Guidelines
Agency Handling	Undersecretariat for Defense Industries (SSM)
Part of Procurement Decision	Yes
Offset Sector	Military
Minimum Value of Contract	US $5 million
Minimum Offset Required (%)	30% of contract, 50% of project import value, will change to 100%
Term	Not defined
Multipliers	1-5 times, based on type of offset
Penalties	10% (on unfulfilled) & temporary exclusion from future bids
Focus	Self-sufficiency, new business opportunities, increase foreign currency flow, improve quality
Direct vs. indirect	Both
Eligible Offset Activities	Exports, technology transfer, R&D, training, investments, etc.
Website	No website available

UAE

Title of Policy:	The Defense and Procurement Policy of the UAE
Agency Handling	UAE Offsets Group (UOG)
Part of Procurement Decision	Yes
Offset Sector	Military
Minimum Value of Contract	US $10 million
Minimum Offset Required (%)	60%

Term	7 years
Multipliers	Yes but unpublished
Penalties	8.5% of offset obligation or 4.5% of total contract
Focus	Sustainable wealth creation
Direct vs. indirect	No distinction
Eligible Offset Activities	Profits of joint ventures with local parties
Website	Offset Program description: http://uae.gov.ae/Government/business.htm#Offsets

United Kingdom

Title of Policy:	Industrial Participation
Agency Handling	MOD/DESO
Part of Procurement Decision	No
Offset Sector	Military
Minimum Value of Contract	£10 million ($16.10 million)
Minimum Offset Required (%)	100%
Term	Over period of procurement contract
Multipliers	0-6 times based on type of offset
Penalties	None, however strict enforcement of IP program
Focus	Provide new
Direct vs. indirect	No distinction
Eligible Offset Activities	Mainly technology transfers, also employment, equipment, purchases and other
Website	"Doing Business with the Ministry of Defence". http://www.dgcom.mod.uk/dgcom/guides/sect05.htm

Appendix B

LIST OF INDUSTRIES FOR WHICH INDUSTRIAL LICENSING IS COMPULSORY

Under Industries (Development & Regulation) Act, 1951

1. Distillation and brewing of alcoholic drinks.

2. Cigars and cigarettes of tobacco and manufactured tobacco substitutes.

3. Electronic aerospace and defence equipment: all types.

4. Industrial explosives including detonating fuses, safety fuses, gun powder, nitrocellulose and matches.

5. Hazardous chemicals:

    a. Hydrocyanic acid and its derivatives

    b. Phosgene and its derivatives

    c. Isocyanates and diisocyanates of hydrocarbon, not elsewhere specified (example: Methyl Isocyanate)

6. Drugs and Pharmaceuticals (according to modified Drug Policy issued in September , 1994 and subsequently amended in February, 1999)

Note: Manufacture of SSI reserved items by other industrial undertakings and location of industrial undertakings in relaxation of the notified locational policy will attract compulsory licensing.

Appendix C

Ministry of Science & Technoloy

Department of Scientific & Industrial Research

Industrial R&D Promotion Programme (IRDP)

Fiscal Incentives

Type of Institutions	Incentives	Notifications/Acts/Orde-rs	Approval & Application Forms
Industries having in-house R & D Centers	i. Write off of revenue expenditure	Section 35(1) (i) of Income Tax Act	
	ii. Write off of capital expenditure	Section 35(1) (iv) of Income Tax Act	No approval required. Cases referred to the prescribe authority (DGIT (E) Calcutta in concurrence with Secretary DSIR) through CBDT, in case of dispute with assessing officer.
	Duty free import of Pharmaceutical reference standards	Notification No. 26/2003-Customs dated 1st March 2003 (item at serial no. 138)	In-house R & D centres should be approved by DSIR. Click for Form (Recognition , Renewal of recognition)
	Duty free import of analytical & specialty equipment List 28 for R & D.	Notification No. 26/2003-Customs dated 1st March, 2003 {item at serial no. 248 (1)}	In-house R & D centres should be approved by DSIR. Click for Form (Recognition , Renewal of recognition, yearly return)
	Duty free import of analytical & specialty equipment List 28 for production	Notification No. 26/2003-Customs dated 1st March, 2003 {item at serial no. 248 (2)}	In-house R & D centres should be approved by DSIR. Click for Form (Recognition , Renewal of recognition, yearly return)

Type of Institutions	Incentives	Notifications/Acts/-Orders	Approval & Application Forms
Industry / Private Sponsored research programmes	iii. Weighted tax deduction of 125%	Section 35(2AA) of the Income Tax Act.	The Head of the concerned National Laboratory or the University or the Indian Institute of Technology.
Companies engaged in business/manufacture/production in following areas - Bio-Technology Drug & Pharmaceuticals Electronic equipment Computers Telecommunication equipment Chemicals Aircraft's and Helicopters NEW Automobiles including Automobile Components	Weighted tax deduction of 150 %	Section 35(2AB) of the Income Tax Act)	Click for Guidelines & Form
Commercial R & D Companies	Ten years tax holiday for companies approved by DSIR before 1st April 2004. NEW Extended from April 1, 2004 to March 31, 2005	Section 80-IB(8A) of Income Tax act	Click for guidelines & application forms
Investments on plant & machinery based on indigenous technology	Accelerated depreciation allowance @ 40 %	Rule 5(2) of Income TaxR ules.	Click for guidelines & forms
Public funded R & D institutions	Duty free import of equipment and consumables. Excise duty exemption on equipment and consumables	Notification No.51/96 - Customs dated 23rd July 1996. Notification No. 10/97-Central Excise dated 1st March 1997.	Click for Form (registration, Renewal of registration) Click for Form

contd/-

Type of Institutions	Incentives	Notifications/Acts/Orders	Approval & Application Forms
Scientific Research Associations with object of undertaking scientific research or university, college or other institutions. University, college or other institutions carrying out research in social sciences or statistical research	Income tax exemption of 125% to donations. Income tax exemption of 125% to donations	Section 35(1) (ii) of Income Tax Act Section 35 (1) (iii) of the Income Tax Act.	Prescribed authority is Central Government Prescribed authority is Central Government
SIROs recognized by DSIR	Duty free import of equipment and consumables Excise duty exemption on equipment and consumables.	Notification No.51/96 - Customs dated 23rd July 1996. Notification No. 10/97- Central Excise dated 1st March 1997	Click for Form. Click for Form.
Any wholly owned Indian company	Excise duty exemption for 3 years on goods designed and developed and patented in any two countries out of: India , USA , Japan and any one country of European Union	Notification No. 15/96 - CE dated July 23, 1996 , amended vide Notification No. 13/99- CE dated February 28, 1999.	
Any Indian company	Exemption from price control of DPCO Revenue expenditure on R & D laid out within three years immediately preceding the commencement of business	Drugs (Prices Control) Order 1995 Section 35 (1) (i) of Income Tax Act	Click for Form. Prescribed authority is DGIT (E) Calcutta in concurrence with Secretary DSIR
R & D projects funded by Government of India	Exemption from Custom duty on imports.	Notification No. 50/96- Customs dated 23 July 1996	

Last Updated: November 25, 2004

Appendix D

HOLDING COMPANY MODEL FOR DEFENCE INDUSTRIAL BASE

The concept of "Holding Company" for management of business enterprises has been in vogue in the western counties and in India for quite some time. As far as public sector in India is concerned, this concept was suggested for the first time by Administrative Reforms commission in 1967. The commission suggested creation of several industrial and business units in key sectors for placement under a holding company in each major sector like steel, coal, petroleum etc. In 1984, the Arjun Sen Gupta Committee suggested holding company form of organization, especially for enterprises in core sector. The committee felt that this would help reduce span of control for the Government by reducing the control points to a limited number of large comp anies. This arrangement would also ensure the required autonomy to enable the functioning of such enterprises on purely business considerations.

During the 1970's and 1980's several holding companies come into existence in India in steel, coal, textile and trading sectors etc. These experiments with holding company in public sector had varying degree of success.

Defence Industrial Base (DIB) in India, mainly consists of a number of organizations, in public sector like HAL, BEL; those under the control of the Government like ordnance factories, DRDO; those under the control of armed forces like rep air workshops etc. The latest entry in this field is the private sector.

By and large, soviet model of management of defence industrial enterprises has been followed by India. China too followed the same model earlier. DIB in India in the Government sector, consists of a number of manufacturing enterprises and a separate R&D set-up, and in many cases sep arate inspection too. For example, Indian Ordnance Factories do not have mandate or facilities for R&D. Knowledge acquired by their manpower during the course of production of armament could he fruit fully used for R&D, if they were also responsible for R&D to bring out new products.

After liberalization of economies in Russia, China and India, competition for defence enterprises has intensified and new opportunities have arisen for export of defence goods and services. There is also a better environment for transfer of technologies from one country to another. To cope up with the new environment, both Russia and China, have gone in for massive reorganization of their defence enterprises. The main objectives being: making these enterprises globally competitive in respect of quality, cost, speed in delivering goods and services, and bringing in new technologies ahead of competition to mention a few. The organization pattern followed earlier was considered inadequate to meet the new objectives. They have progressively gone in for a "Holding Comp any Structure" to make their enterprises globally competitive. This new structure has been described earlier while dealing with Russian and Chinese defence industries.

In India, we are, by and large, still persisting with the old organization structures. We have not succeeded in making our DIB globally competitive. We continue to import nearly 70% of requirement for defence industrial goods. There are complaints that the cost of domestically produced goods in high, quality is not competitive and speed of delivery does not meet the requirement of the armed forces. W e do develop new technologies, but in many cases the commercialization is so late that it results in imports. The effort put in by the R & D setup is more or less wasted, mainly because we do not have suitable organization structure to reap the benefits of R & D.

In India too we could consider forming holding companies for various products required by the armed forces. A holding company, like in China and Russia, will have all the activities under their control to bring out and service the end product, including production, R&D, marketing, quality assurance, finance, human resource management etc. With one entity being responsible for everything connected with a product, need for co-ordination among various authorities would go down, thereby improving the decision making process, faster commercialization of new technologies, better sharing of experience within the organization and many other benefits. Eventually when D I B expands, more than one holding company could be formed for a set of products to bring in an element of competition, as being done by China.

Carrying out this kind of re-organization is not easy , but it is desirable if by the year 2025, we have to be self-reliant for our requirement of arms to the extent of seventy percent or more.

Appendix E

Press Information Bureau Government of India

Friday, February 27, 2009

Ministry of Science & Technology

NEW MILLENNIUM INDIAN TECHNOLOGY LEADERSHIP INITIATIVE (NMITLI) SCHEME

19:10 IST

As a part of New Millennium initiative, the Government mounted a farsighted R&D Programme named 'New Millennium Indian Technology Leadership Initiative (NMITLI)' in Public-Private Partnership mode in 2000-01. The programme was announced as part of the Union Budget in the year 2000. The responsibility of conceptualizing, evolving and implementing the programme has been assigned to the Council of Scientific & Industrial Research (CSIR).

The trigger for NMITLI programme was:

- From incremental innovation to disruptive innovation;
- Tolerance for risk taking and failure;
- Best minds in India to take up the grand challenge for collaborative excellence; and
- Technology leadership.

The NMITLI focus is to:

- identify niche areas where India can gain leadership in about 10-15 years;
- develop projects involving best brains of the country through a rigorous process;
- build knowledge network of partners from public funded institutions and private industries;
- develop new methods of working together for collaborative excellence;

- focus on proof-of-concept; and
- provide a pipeline of cutting edge Indian innovation for conventional technology financing bodies as against the 'usual safe bets.

Today, the New Millennium Indian Technology Leadership Initiative (NMITLI) is the largest public-private-partnership effort within the R&D domain in the countryIt looks beyond today's technology and thus seeks to build, capture and retain for India a leadership position by synergising the best competencies of publicly funded R&D institutions, academia and private industry. The Government finances and plays a cat alytic role. It is based on the premise of consciously and deliberately identifying, selecting and supporting potential winners. NMITLI has carved out a unique niche in the innovation sp ace and enjoys an excellent reputation.

NMITLI has so far evolved 57 largely networked projects in diverse areas vizAgriculture & Plant Biotechnology, General Biotechnology, Bioinformatics, Drugs & Pharmaceuticals, Chemicals, Materials, Information and Communication Technology and Energy . These projects involve 80 industry p artners & 270 R&D groups from dif ferent institutions. Approximately 1700 researchers are engaged in these project s. These 57 project s cumulatively have had an outlay of approximately Rs. 500 crore.

NMITLI Achievements:

The programme has generated about 100 international p atents and 150 publications in peer reviewed journals. The important achievements are:

- Paradigm shift in leather processing-From chemical to biochemical route
- Pilot plants for separating cellulose, hemi-cellulose and lignin from bagasse
- Pilot plants for producing lactic acid from sugarcane juice
- Bio-informatics software viz. Bio-Suite, GenoCluster, Bio-SuiteC and Darshee
- Developed three variants of SofComp (Simple ofice Computer) devices including Mobilis
- Anti-psoriasis formulation in Phase-III Clinical Trial
- Lysostaphin in Phase-II Clinical Trial
- Anti-tuberculosis molecule in Phase-II Clinical Trial
- poly herbal formulations for diabetes, arthritis and hepatic disorder

- Micro-PCR based immuno-diagnostics for detecting eye infections
- Development of new plant varieties of Mentha piperita
- Development of Triple-Play broadband technology

Key components of CCEA Note: Enthused by the success of the programme and on the recommendations of several committees, Government has approved the expansion of NMITLI programme to experiment newer models of innovation development. The key components of the proposal are:

Funding along with industry (50:50 Initiative) There are many Indian companies who are doing financially very well but do not have the necessary expertise and intellectual resources to develop focused network projects for development of technologies/products in their line of activities.Their efforts need complementation from suitable R&D institutions and guidance from recognized peers to develop and commercialize newer technologies/ products. Therefore, NMITLI will leverage its experiential base to encourage and assist such companies for developing network projects for those companies in product/technology development through a specific scheme called 'NMITLI 50:50 initiative'.

Co-financing with Venture Capital funds Many venture capitals are limited in scope and risk taking, due to lack of domain knowledge within the organization. Venture Capitals are therefore interested in joining hands with NMITLI, which has strong domain knowledge base, to jointly finance projects. Such projects would be identified and evolved following the procedures established by NMITLI. The funding would be joint with pre-determined ratio, but not more than 50% contribution from NMITLI.These projects are envisaged to be monitored by a joint team of experts as per the NMITLI monitoring mechanism.The proposed funding would follow the venture funding norms.The successes and failures resulting from the projects will be shared on equitable basis.

Setting up of NMITLI innovation centres in selected areas for long term sustained efforts Some areas need long term sustained support with requisite human resource as well as infrastructure, assembled at one place to cross the threshold of intellectual barrier in order to generate globally competitive technologies and products, IPR, and high quality publications. It is envisaged to set up 'NMITLI Innovation Centres' in PPP mode for sustained efforts in some selected areas for example, Photovoltaics, Fuel Cells, White LEDs, Industrial Enzymes, Medical Implants, Vaccine development, Seed Development etc.

Support to post NMITLI projects Despite the excellent R&D and developments, the technologies and products developed in the laboratory do need market seeding, pilot plant studies to refine the development. The companies need CSIR's hand holding to develop and package the technologies/products further. The concept of 'Post-NMITLI' will fulfill the objective of providing financial and technical assistance for pre-commercialization related activities such as scale up, pilot plants, field trials, market seeding of products, market surveys, etc.

Acquisition of early stage relevant knowledge / IP for portfolio building. External ideas / leads / IP acquisition are assuming greater significance in the chain of innovation and mind to market. The availability of a large number of unencumbered IP (being developed in several laboratories globally) is providing a fillip to this approach. Several countries across the globe are striving to take advantage of the diversity of creativity available in different parts of world and integrate with its own developments to bring out new products / processes for global competitiveness. Since NMITLI aims to provide the Technological Leadership to the Indian industry it becomes imperative for NMITLI to adopt such practices to achieve its objectives. Such acquisitions shall be in chosen areas with a view to creating a portfolio where NMITLI projects are in operation.

Crossing the geographical boundaries It is increasingly being felt that to achieve leadership in niche technology areas, relying totally on internal expertise and capabilities may not be adequate. To achieve the objective of global leadership, it would be helpful to broaden the programme by bringing in international expertise. The international expertise may be in the form of expert advise of international experts at various stages of project development and implementation, involving international companies for product/technology development and commercialization at global scale, and engaging research institutions and/or CROs across the globe where Indian expertise need outside complementation.

Joint development and support of projects with other departments of science and technology as well as economic ministries Many government departments are engaged in research and development activities in areas of relevance to them. These activities often have considerable degree of overlap with other scientific departments. However, these departments's expertise is limited to undertake multi-disciplinary projects in cutting edge areas requiring wide-spectrum of intellectual and infrastructural inputs. Such multi-disciplinary areas need expertise, inputs and concerted efforts from all concerned government departments to generate IPR, technologies and products besides high quality publications. Therefore, part of the NMITLI funds will be utilized to generate inter-departmental projects in the XI FY Plan. The proposed scheme apart from generating intellectual capital, technologies and products in cutting edge areas would act as a catalyst to bring better co-ordination among various departments of government in the R&D sphere.

Relaxing the condition of more than 50% shareholding by Indians/Non-resident Indians Many oversees companies through their R&D efforts using local resources, produce goods for local as well as overseas consumption and are thus contributing to the growth of Indian economy. They also employ Indian workforce. In some areas, such companies are better equipped to upscale the technology/product s and sell it under their brand name. Further, they can become a vehicle for taking Indian technologies and products into global market easily thereby contributing more to Indian economy.The provision permits relaxation of the condition of more than 50% shareholding by Indians/Non-resident Indians to become an industrial partner in NMITLI projects.

Flexibility to convert loan into equity Launching a new product or setting a knowledge based new venture requires investment on many front s particularly for capital-intensive infrastructure, manpower, technology costs, working capital etc. Governments all over the world, particularly in developed countries endeavor to support entrepreneurs in different ways to ease the burden of initial investment. The industrial partner under NMITLI has to invest for commercialization of technology/products and at the same time has to return the loan to CSIR albeit in installments. This burden of loan repayment can be further reduced by converting loan into equity Therefore, with this provision and on the request of industrial partner, loan given to it may be converted into equity.

Advantages of NMITLI Expansion: As others are emulating, NMITLI is endeavoring to position higher in the innovation development. The proposal will therefore:-

- Enable CSIR to experiment newer models of innovation development in Public-Private-Partnership (PPP) mode, which could later become models for others to emulate;
- Encourage to develop product s / processes based on innovation and thereby help Indian industry emerge as a technology leader in the identified domain;
- Encourage venture capital funds to venture into more risky R&D areas;
- Act as a catalyst to bring better co-ordination among various departments of government in the R&D sphere and avoid unnecessary duplication, apart from generating intellectual capital, technologies and product s in cutting edge areas; and
- Enhance national competitiveness.

PRA/SKK

www.ingramcontent.com/pod-product-compliance
Lightning Source LLC
Chambersburg PA
CBHW080933240926
51849CB00003BA/449
* 9 7 8 9 3 8 0 1 7 7 3 4 2 *